RECORDER

Recorder Technique

INTERMEDIATE TO ADVANCED

ANTHONY ROWLAND-JONES

Second Edition

Oxford   New York
OXFORD UNIVERSITY PRESS
1986

Oxford University Press, Walton Street, Oxford OX2 6DP

Oxford New York Toronto
Delhi Bombay Calcutta Madras Karachi
Petaling Jaya Singapore Hong Kong Tokyo
Nairobi Dar es Salaam Cape Town
Melbourne Auckland

and associated companies in
Beirut Berlin Ibadan Nicosia

Oxford is a trade mark of Oxford University Press

Published in the United States
by Oxford University Press, New York

First published 1959
Second Edition 1986

British Library Cataloguing in Publication Data

Rowland-Jones, Anthony
Recorder technique: intermediate to
advanced.——2nd ed.——(Instrumental
technique books)
1. Recorder (Musical instrument)
I. Title   II. Series
788'.53'0712   MT350
ISBN 0-19-322342-2

Library of Congress Cataloging-in-Publication Data

Rowland-Jones, Anthony.
Recorder technique.

Bibliography: p.
Includes index.
1. Recorder (Musical instrument)——Methods.
I. Title.
MT351.R72   1986   788'.53'0714   86-12442
ISBN 0-19-322342-2 (pbk.)

Printed in Great Britain by
J. W. Arrowsmith Ltd., Bristol

To Christina, my best pupil

Contents

Introduction

Since *Recorder Technique* was first published in 1959 a most
gratifying change has taken place in the status of the
instrument, partly associated with the greatly increased interest
in early music generally. Moreover, the recorder has been
discovered by avant-garde composers to be capable of a
diversity of interesting sounds. As an instrument that is easy to
learn in the early stages, it remains a mainstay of music in
schools. Yet the considerable technique required to realize its
capabilities in full, and to respond to the interpretive difficulties
of much music written from the sixteenth century onwards, has
given reason for musical academies throughout the western
world to offer degree courses specializing in the instrument. The
recorder, unlike the flute, has not fundamentally changed since
medieval times, but it has developed in response to changing
demands of musical style: the desire to achieve authenticity in
recorder playing in all periods thus stimulates a process of
interesting research and discovery. The double challenge of
virtuosity and scholarship has led to the emergence of
professional recorder-players who have demonstrated to other
musicians and to the musical public that the recorder is far more
than a mere toy or school instrument.

An enormous increase in the number of serious amateur
recorder-players has encouraged music publishers to extend
their catalogues of music suitable for the recorder, and some
new publishers specialize in early music. The original chapter
in *Recorder Technique* on the recorder's repertoire, even with its
appendix in editions from 1969, is now considerably out of date.
It would not in fact be practicable in a short space to attempt to
review the published repertoire with any degree of complete-
ness. In this revised edition, repertoire is presented more
selectively (Appendix 1): even so, important new editions or
compositions may appear between the time of writing and
publication of this book. Similar factors apply to the list of
makes of recorders, a subject which would now qualify for a

book in itself were it not for the happy circumstance that new names so frequently appear on the scene as to make such a project impracticable. Objective information about makes of recorders can best be obtained from specialist dealers such as Saunders Recorders in Bristol.

The increased interest in the recorder has also led to what has been termed in the pages of *The Recorder* (Schott), the journal of the Society of Recorder Players, as an 'information explosion'. This has been manifested not only through that journal, but through many articles in *Early Music* (Oxford), and in other music journals and books (see selected bibliography, Appendix 2). Major works of scholarship have been published on the interpretation of early music (e.g. by Donington), and especially on ornamentation (e.g. Neumann). Reprints of recorder tutors from that of Ganassi (1535) onwards, are now available, together with new books reflecting the methodology of eighteenth-century tutors (see for example under Veilhan in Appendix 2). The nine-page history of the recorder in the original *Recorder Technique* was superseded in 1962 by Edgar Hunt's *The Recorder and its Music* (Herbert Jenkins). An indication of change is that Mr Hunt (p. 95) then said 'today it would seem that musicians in Italy think only of opera, singing and strings'; nowadays there is a recorder festival in Urbino and an Accademia del Flauto Dolce in Turin. Of particular relevance to this much revised edition of *Recorder Technique* was the appearance in 1962 of my *Practice Book for the Treble Recorder* (Oxford) and in 1978 my *Introduction to the Recorder* (Oxford). The latter renders superfluous the chapter in the original *Recorder Technique* on beginning to play the recorder.

This book in its revised version, therefore, concentrates on recorder technique for the player who wants to advance from the first stage of recorder playing to the point at which he may be termed a 'good amateur'. No book can do this on its own, however, and it is essential that individual coaching should be sought from a good recorder teacher. A player who has worked through *Introduction to the Recorder* will, with experience, be able to play most recorder consort music and at least to attempt a Handel sonata. *Recorder Technique* will help him to achieve good public performances of Handel and other sonatas and to play chamber music alongside professional players. It does not deal with the high levels of technique and scholarship required to give professional performances of Berio's *Gesti* or an ornamented improvised recorder part of a fifteenth-century *basse danse*. Nor does it expect its readers to practise scales and *passaggi* several hours daily, as a professional player must.

It does, however, in its absence of musical examples, expect its readers to have a treble recorder to hand to try out many suggestions (e.g. of fingerings). It would be a considerable advantage to relate its chapters to the relevant sections of the *Practice Book* (*PB*) to which cross-references are made. A player who, with his teacher's help, masters the material in *Recorder Technique* and demonstrates this through playing the extracts in the *Practice Book* (and longer compositions of a similar standard) both accurately and musically has achieved the technical ability expected of a final degree student, and should honourably call himself a 'good amateur'.

I
KNOWING YOUR INSTRUMENT

There comes a point in the process of learning to play the recorder where, in a sense, you have to go back to the beginning. One can draw an analogy with driving a car: by passing a test and gaining experience you can drive a car safely and adequately, but to drive a car *well* involves knowing its mechanism and considering its performance under pressure. A recorder-player, however skilful his technique, cannot play his best unless he knows his instrument, its qualities, strengths, and weaknesses. This opening chapter is intended to increase your knowledge about recorders generally, and to help you establish a rapport with your instrument, a prerequisite to good performance.

A recorder is basically an open tube blown at one end in which sound is produced by the impinging of the air-stream shaped by a windway upon an edge formed by the chamfering of the surface of the tube. This principle of sound formation is used for the flute, the only difference being that a flute-player shapes the air-stream with his lips. A page of this book can be made to produce a convincing edge-tone: hold the book open with both hands so that one page is separated and bring the edge of the page to about a quarter of an inch from the lips, keeping the page in a horizontal position. Now blow softly a very thin stream of air directly on to the edge, and, with adjustment, a quiet wavering squeak will result. This squeak is caused by the formation of eddies on each side of the page which produce regular alternations of pressure and set the air around vibrating (for further details and diagrams consult books such as *Music and Sound* by L. S. Lloyd, Oxford, 1951). In the flute and the recorder these vibrations are modified by being 'coupled' to a tube the sounding length of which may be altered by opening and closing holes in it with the fingers and thumb. The recorder uses eight such holes, although the two lower holes are doubled to facilitate the production of semitones, at least in modern and late baroque instruments.

Experiment further with the edge of the page by blowing even more softly and holding the page a little more away from the mouth. This will produce a very quiet edge-tone of low enough pitch to be identified on the piano, illustrating the principle that the softer one blows the lower the note, and vice versa. This principle is used to obtain high notes on the flageolet (which normally has six finger-holes), and on the tabor-pipe, which, with its long, narrow bore susceptible to harmonics, can produce two octaves from three holes by the use of stronger articulations and overblowing. Unless the cross-sectional area of the windway can be controlled (as in the lips of the flute-player), overblowing means more air, and more air creates more sound, so on the flageolet the upper octave must needs be louder than the lower. The formation of harmonics on a wind instrument can, however, be achieved without any increase of breath-pressure by opening a small hole (octave-hole or 'speaker') between the source of vibration and the first open finger-hole: on the recorder this is usually done by partially opening the thumb-hole. The superiority of the recorder over other end-blown flutes lies in its ability to play notes of the upper octave more softly than those of the lower octave. Recorder-players must exploit this facility to the full in response to the needs of the music they play.

The windway of a recorder is formed by the insertion of an incised plug or block (whence the German name 'Blockflöte') into the blown end of the tube, which is generally made beak-shaped (whence 'Flûte à bec') to fit between the lips. This block is made of a wood such as red cedar which does not swell with moistening. The opening of the windway opposite the edge is chamfered slightly to direct the air in the best way against the edge: this process is called 'voicing'.

It is in the skill and understanding with which the craftsman matches the voicing to the nature of the piece of wood he is working with, and to its future maturing as a musical instrument, that the handmade wood recorder is so superior to a machine-made instrument, although misjudgement or bad luck could result in a handmade recorder being a very poor thing. The most critical factor in voicing is in the way the edge divides the air into two parts, one vibrating in the bore, the other escaping along the top of the instrument. An uneven division of air favours the tone of the lower notes of the instrument, and minimizes differences in quality between the notes of this register, but this voicing makes it difficult to get good high notes, which are better produced with an equal parting of the air-stream. The distance between the opening of the windway and the edge is also a factor in tone production, the higher register being favoured by closeness, although too close a voicing stifles the tone: if this distance is made greater the lower notes are favoured, although too great a distance means that less air is made to sound, so causing fluffiness. The cross-sectional area and profile of the windway affects tone because this determines the amount of breath needed to make a note. Many other facets of construction affect tone and volume, such as the height and angle of the window walls, the breadth or narrowness of the window, the shape of the windway (parallel or converging), the thickness of the walls of the bore, the undercutting of the finger-holes, and, especially, the width and profile of the bore – whether it is round or oval, whether it is cylindrical or conical, or (as with most modern recorders) conical at first and then opening out slightly up to the bell.

It is important to consider these factors, because the maker of the instrument will have had a particular tone-quality in mind in constructing his instrument, and a player is at odds with his instrument if he does not identify and nurture this tone. The best tone may be achieved with lowish breath-pressures, or, if the windway opening is narrow, with a rather higher breath-pressure. The player must experiment to know where the instrument responds best across its range. If the recorder is properly looked after (see below) tone may improve with age, but it will deteriorate when the thumb-hole becomes worn and needs re-bushing, or when the edge and chamfers need sharpening, and revoicing by the maker is required.

Although voicing in relation to bore is paramount, tone-qualities are also associated with the material of which the instrument is made. As Bacon puts it, 'When the sound is created between the blast of the mouth and the air of the pipe, it hath nevertheless some communication with the matter of the sides of the pipe, and the spirits in them contained' (Natural History, cent. II § 167). Wood is undoubtedly the most satisfactory material for making recorders. Bacon suggested that 'it were good to try recorders and hunters' horns of brass, what the sound would be' (cent. III § 234). The effect would in fact not be pleasant, since brass has a ringing tone of its own which would react favourably to some notes and unfavourably to others, resulting in inequality and hardness. Ivory produces an elegant but hard tone-quality. Despite their weight, and sometimes (even after long seasoning and careful selection) a tendency to crack, very hard woods make the best recorders. They can be voiced with sharp, firm, long-lasting edges; softer woods deteriorate more quickly, and absorb or dampen sound. The best wood for recorders has no knots and a close, parallel grain that allows a surface to stay smooth even under conditions of frequent wetting and drying, without cracking, splintering or swelling. Even microscopic splintering or misshapenness at the windway, edge, bore, or finger-holes will cause imperfections of tone. The woods that most nearly meet these requirements are expensive hardwoods, such as box, palisander, rosewood, cocobolo, olive, and African blackwood. Slightly softer woods such as maple, plum, pear, and cherry are also suitable for recorder making (for types of wood used in making recorders see Hildemarie Peter's book and Michael Zadro's essay in *Early Music*, cited in Appendix 2). Woods such as maple can be made impermeable to moisture by heat treatment followed by impregnation with paraffin wax. Such instruments are more stable and less likely to crack, but they are somehow less satisfying to play upon than naturally seasoned wood with its greater individuality and vitality. Although one may have a preference towards, say, box or rosewood, some makers assert that the hardwoods listed above do not each produce a characteristic tone-quality, i.e. that the differences in tone between instruments made by the same maker depend entirely on the subtle relationship between voicing and the nature of the piece of wood used for each instrument. Other makers, however, feel that there is a generic difference between all their instruments made from, say, box (the eighteenth-century favourite, though it is liable to crack) and, say, rosewood (a modern favourite).

A further vital factor affecting a player's approach to his instrument is the fingering system for which the instrument is designed. With the demise of 'German fingering' (0 123 4––– for B♭ but 0 123 –567 for B♮) modern recorders are all now constructed to achieve good intonation with the first and second octave normal fingerings familiar to readers of this book (see inside back cover of *Introduction to the Recorder* or the fingering chart provided with a new recorder). This is not necessarily the case with recorders which are replicas of renaissance instruments or of even some baroque instruments (see below).

'Normal' fingerings are in any case as much a compromise as equal temperament in keyboard tuning, and makers will expect players beyond the elementary level to accommodate intonation to changes of key. For example, normal C♯ may be good as the leading note in D major, but slightly sharp as the major third in A major. Moreover, the instrument maker may not have succeeded in the extremely difficult task of getting all the normal fingerings true to equal temperament (if that is what he is aiming at): an 'alternative fingering' then has to become the normal fingering for that note. For each instrument the best fingering for each note must be discovered: hopefully it will accord with normal fingering for most circumstances. Notes above top F″ are liable to need a particular fingering for each recorder. This is the case especially with bass recorders. Unusual fingerings required for each instrument should be noted on a card and kept with the recorder for reference. These considerations apply not only to intonation but also to tone. A note may be in tune with normal fingering but poor or obtrusive in tone-quality. Such a problem may very likely be cured by a change in tonguing (usually to lighter tonguing) for that note, or by a change in breath-pressure, but occasionally the player is better off by always using an alternative fingering for the bad note. This often applies to bass instruments; and small recorders such as the sopranino in F and the little garklein in C are difficult to construct absolutely in tune, although intonation and tonal faults are less noticeable on high-pitched instruments.

Having to make corrections for intonation or tone may be tolerable on bass and high instruments, but you would be overburdening yourself if you had to remember continually to make substantial adjustments on your main solo instruments. The maker might be prepared to re-tune the instrument if you are absolutely certain that it is a poor one (and you need to get a reliable second opinion about this). Or you can, if you are sure of yourself both in ear and hand, re-tune your own instrument (see references in Appendix 2). Otherwise you would do better to obtain another instrument rather than engage in a constant struggle against odds. But it is the case that many players have to go through a love–hate relationship with their instrument before coming to terms with it. Assuming that their instrument is reasonably good, they at first accept what it tells them in response to 'normal' fingering. As they get beyond the first stage of recorder playing, they become more critical and make more demands on the expressiveness of their instrument. Then they begin to realize imperfections: possibly their own sense of intonation improves (although the opposite happens to some players who become convinced that their recorder's high E' is in tune when it is not). This is the point at which advice from an experienced player is essential, as to whether to persist (which is usually the case) or whether to change instruments (rarely). They then discover, or are shown, how to put things right, sometimes by quite small modifications in breath-pressure or tonguing, without re-fingering. Finally, these modifications become instinctive, and players will automatically make such modifications upon handling a particular recorder, without consciously remembering what each instrument needs. This is an important part of recorder playing, for no recorder is perfect.

The qualities of recorders vary considerably from one maker to another, since the variables in designing and building a recorder are enormous. This is why consorts of instruments by one maker often sound better than consorts of instruments by different makers. Yet the same maker making two treble recorders in the same material is likely to produce two different-sounding instruments. Even plastic machine-made recorders by the same producer contrive to have individual qualities. It is one of the joys, as well as one of the problems, of recorder-players that instruments are so different.

Historical differences in recorders

Most recorders available in shops are based on principles of design not much different from eighteenth-century models, so-called 'modern baroque'. They are available in six sizes in F and C – at modern orchestral pitch A = 440 (though modern pitch in practice is creeping higher). This should be 'warmed up' pitch, and not the flatter note produced when a recorder is cold (for example in a shop). Eighteenth-century pitch was a semitone lower, and makers today are making instruments at 'low pitch' A = 415 – although even that differs from many historical originals: but this is the pitch generally accepted for 'authentic' performances of baroque music (see Fred Morgan's essay in *Early Music*, cited in Appendix 2). Woodwind instruments generally produce a richer, deeper, more rounded tone at low pitches, while violins and pianos become more brilliant and louder at higher pitches. A modern recorder will never sound quite like a baroque recorder because of this pitch difference, and for other reasons.

While modern recorders have F and C as their lowest notes, renaissance, and possibly also medieval recorders, were often pitched in A and D, at least down to tenor (in C). Praetorius (1619) refers to a bass in Bb; and several other sizes can be obtained from modern makers. In the eighteenth century, music was written for the sixth-flute (descant in D) and voice-flute (treble/alto in D). Although such instruments in principle require no special techniques not covered in this book, they present challenges in reading or transposition which a fully-trained recorder-player must eventually master (see Veilhan's book on technique, cited in Appendix 2). Musicians in the eighteenth century, and more so in earlier periods, must have been adept at reading in many different clefs, and renaissance musicians would have abhorred our multiple leger lines.

Baroque recorders

Apart from pitch, and the wider range of instruments available, the baroque recorder varies from 'modern baroque' in its voicing and fingerings. The reforms made by the Hotteterre family in the late seventeenth century resulted in recorders being made with a narrower, double-conical bore (tapering, then widening). This in itself favoured high notes and increased the compass to cover the high ranges much used by Bach and Telemann (e.g. the second movement of the second Brandenburg Concerto – *PB* Ex. 45). The Hotteterres first produced the recorder in three parts, with strengthening at the two joints being both a necessity and an ornament – the bulges and rings which delightfully reflect the decorative taste of the time. The three-section recorder enabled makers to use shorter pieces of wood, aiding selection and accuracy of internal finish. Voicing, too, favoured the higher partials in the position of the edge in relation to the windway opening. The characteristic of baroque recorders most affecting technique, however, is the vertical narrowness of the windway exit, which offers greater resistance to breath-pressure. This causes less air to be used in making a note, resulting in less sound. Makers partly compensated for this by widening the breadth of the opening and edge, which required both to be arc-shaped, matching the exterior curve of the head section and avoiding leaving too little thickness of wood at the sides of the windway, which, even so, was often protected by encasing in ivory. Players likewise compensated by using higher breath-pressure. There are two great advantages in this design. The first is that it is less difficult (less 'clickish') to cross register breaks. The second, even more important, is that this increases *flexibility*, i.e. a note can be played louder or softer without going out of tune, so increasing the expressive range with less recourse to some of the intonation-control techniques discussed later in this book. Baroque instrumental music, like baroque opera, is highly expressive and emotional, and even on a baroque recorder the volume range that should be used is beyond the flexibility available in using only normal fingerings. A baroque-designed instrument is ultimately therefore no easier to play than a 'modern baroque' recorder, but its flexibility encourages a style of playing that comes less naturally on the modern recorder.

The baroque recorder tends to be rather reedier in tone than its modern derivative, which has a more open tone. The volume is likely to be less than a modern recorder at full breath-pressure, although volume in itself is not always an advantage – what matters is penetration of tone. The modern recorder has many advantages if it can give a good tone-quality at low breath-pressure, and yet, with shading, a strong tone at high pressure. What is lost in flexibility is gained in variety. On the purely practical side, the modern design with the vertically wide windway exit, apart from being easier to make, does not have such a propensity to clog up with moisture as the baroque voicing. A baroque recorder needs constant care to prevent fluffiness, particularly on those all-important high notes which it is designed to play so beautifully. Professional players, most of whom naturally choose the baroque recorder to play baroque music, may need to have a second treble recorder ready and warmed up for use in the second part of a recital because of the danger of clogging up, particularly if the atmosphere is humid.

Many recorder makers now make 'baroque' recorders based closely on eighteenth-century models, with baroque voicing. It is vital for the player to know how his recorder is voiced, or he could play all the while at the wrong breath-pressure. These makers do not necessarily, however, copy baroque fingering. The first difference is that baroque makers up to the time of Bressan (*fl.* 1685–1731) seem rarely to have used double holes on the bottom two fingerings, although Hotteterre (1707) mentions their availability. This is because half-holing techniques during the Renaissance and baroque periods were much more widely used on all fingers than we use nowadays. Modern makers of baroque recorders are usually prepared to provide double holes.

The other fingering differences are rarely copied, for players and makers generally agree that they are a disadvantage. Hotteterre's fingering chart shows finger 6 down for most notes, presumably copied from normal technique on the one-keyed flute. This is partly to support the instrument (baroque instruments do not have the modern advantage of a thumb-rest – see below), but the instrument would have been built to be in tune with that fingering. Later baroque fingering charts show Bb as 0 123 4–6–, i.e. without the little finger, whereas modern fingering, and that used by most makers of baroque instruments, has the little finger down for low Bb, facilitating the intonation of B♮ and high Bb'.

Renaissance recorders

The greatly increased interest in renaissance music, especially instrumental music, has given an impetus to the production of recorders based on sixteenth-century models and illustrations. The outstanding differences between baroque and renaissance recorders is that the latter are usually built in one piece (but often copied with one or even two joints), and have a wider, more cylindrical bore (see Virdung's illustrations of 1511). This, together with the voicing, facilitates the lower partials, but reduces range – though high notes of dubious quality can still be discovered. The tone-quality is open, not reedy, and the volume is 'beefy' and loud, more obviously so in the lowest notes. Sometimes the tone may sound a trifle coarse, or, with the lower instruments, 'hairy', but these characteristics are less noticeable in ensemble playing – recorders were not generally considered to be solo instruments in the Renaissance period. Yet the voicing used in modern copies allows a sweet tonal response at lower breath-pressure, perhaps because sixteenth-century writers demand that recorder-players should respond to different moods of music as much as if they were singers and the music had words. At higher levels of tonguing, renaissance recorders tend to 'chiff', giving a percussive articulation to each note: it is uncertain how common this usage was, though it sounds most authentic in renaissance dance music. It can easily be avoided by using light tonguings, and, as we can see from Silvestro Ganassi's recorder tutor *Fontegara* (1535), these were even more strongly advocated in the Renaissance than they were in later periods.

Copies of renaissance recorders do to some extent use historical fingerings, and these are mentioned in the chapters on fingering in this book. The main differences are in E♭, F♯', G' (though not always), C♯', and D'.

On the sixteenth-century wide-bore recorder, E♭' or E' were regarded as the highest notes, and it is impossible to get F'', though G'' is present. Some instruments, however, must have been made with slightly narrower, more conical bores, as Ganassi found fingerings for two and a half octaves. In the early seventeenth century, top C'' on the descant was evidently expected to be available, as the solo variations by Van Eyck (see Appendix 2) frequently require it. Models of renaissance recorders are made with the narrower, more conical bore which provides this note, but with a less extrovert renaissance sound. Perhaps this sound, which is pure, round, and rather dispassionate, and one ideal for consort music, should be thought of as the early seventeenth-century sound.

Because they require more air to fill the wide bore, renaissance recorders will at first make the player feel breathless. It is therefore necessary to fill the lungs more deliberately, and to look for more phrasing points in the music at which one can take a breath if needed. This is especially true of bass instruments, which almost seem to swallow the player.

Renaissance pitch was not greatly different from modern pitch, although it was far less standardized. Recorders are therefore made at A = 440.

Renaissance tenor and bass recorders have keys for the bottom note with the keywork protected by a perforated wooden cover called a 'fontanelle'. This can be removed when the keywork needs servicing. It does mean, however, that bottom C♯ (F♯) is missing. These instruments are also heavier, and differently balanced, in comparison with baroque recorders. Support becomes a problem, for the tenor in particular, and a thumb-rest, though desirable, is not authentic. If the bass flute is an example, renaissance wind players seemed quite prepared to put up with some physical discomfort as part of their playing. This somehow fits in with the outgoing quality of renaissance recorders, yet, on the other hand, *squisitezza* (exquisiteness) was equally demanded of the renaissance player.

Playing renaissance instruments in renaissance music calls for a different mental approach, which ultimately can only be achieved by developing a historical sense, and by listening to and understanding music of the period.

Medieval recorders[1]

The oldest extant recorder, in the Gemeente museum at The Hague, was discovered under a house in Dordrecht, and is generally believed to be of mid-fifteenth-century origin, although there is evidence to suggest it may be one or even two hundred years older. Iconography is of doubtful value in relation to recorders because illuminated manuscripts and stained glass generally represent the player in a straight-on stance, so hiding the important evidence of the thumb-hole. The evidence indicates, however, that instruments of the same family came in a wider range of shapes than they did even in the Renaissance period. The Dordrecht recorder cannot therefore be taken as representative. In addition, it is reasonably certain that the recorder consort consisted of only three sizes of recorder – descant, two 'means' or altos, and tenor, possibly in G, C, and F, fairly close to modern pitch, or sharper. The tessitura of the medieval consort is high and piping, in contrast with the deep tones favoured in the sixteenth century.

I am indebted to Brian Carlick of Charlton on Otmoor, Oxford, who makes copies of the Dordrecht recorder (naturally

[1]See also Horace Fitzpatrick's essay in *Early Music*, cited in Appendix 2.

based on suppositions, for the original instrument lacks some kind of head-cap and some kind of foot or bell), for the information that the instrument has a quiet, thin, but sweet tone, with a range of one and a half octaves. It has a narrow cylindrical bore, a narrow mouth, and a more elongated tongue than most later recorders. It is probable that wider-bore recorders with a richer, louder, and very open extrovert tone existed in medieval times. As they were probably always cylindrical (though perhaps with a flared foot) their useful range would not have exceeded an octave and a sixth, like parts in most medieval music. Fingerings for the second octave notes have to be discovered, for a cylindrical instrument does not have the sophisticated tuning of a baroque recorder. Models of medieval instruments, like some renaissance recorders, may at first sound breathy, but they need to be judged in consort or accompanying the voice.

Characteristics of recorders other than descant or treble

A reader of this book who has followed all the advice in *Introduction to the Recorder* will probably possess a treble (alto) recorder in F and a descant (soprano) recorder in C. To achieve versatility in recorder-playing, he must extend this range. This section therefore identifies problems involved in playing the other sizes of recorders generally available, and is followed by a section on choosing a recorder which can be applied to buying any new recorder. It is quite usual at this stage for a player to want a better treble or descant than the cheap one he may have started with.

Tenor recorders

A descant player who also plays treble will have little difficulty in playing the tenor recorder. In fact the tenor is easier to play than the descant. The chief importance of both instruments is as members of the recorder consort – but while there is a substantial repertoire of music in three or more parts without descant, very few consorts are without tenor. The descant has a small solo repertoire of its own, including Van Eyck, and the eighteenth-century sixth-flute concertos (for descant in D); the tenor recorder is much softer, and its tone merges with other instruments, making it of limited value as a solo instrument with keyboard or strings (though it sounds well with guitar). Although the main role of both instruments is in consort or other early music groups, the descant is harder to manage; it is often required to lead, but it needs to be played at low breath-pressure to avoid coarseness, shrillness, and imbalance in the ensemble, and yet remain expressive and under complete control. The tenor's subdued and creamy tone is much easier to achieve, and it is an easier instrument to control. If a player's interests were limited to consort music, he would be as well rewarded in playing only the tenor, as in playing only the treble.

The one problem encountered with the tenor, especially for players with small hands, is in stretching the fingers to cover the holes. Some makers offset the third hole to the left of centre, which helps to solve the problem, though it mars the beauty of the instrument. Many provide a key for bottom C. This has the disadvantage that, unless a double key is provided, C♯ is missing. It also makes intonation control with the little finger difficult to manage. It is better to try to stretch, and become used to it, than to have a C key.

Tenor recorders tend to lack clarity of tone on their highest notes, although these are not greatly used in consort music. Attempts were made by Thomas Stanesby about 1732 to improve and revive the tenor (with its advantage of being in C like flutes and oboes) to challenge the treble as the main solo recorder; but the treble size of recorder gives the best compromise between pungency and sweetness of tone, between carrying power and blending with other instruments, and between richness in the lowest notes and delicacy in the highest. It is important for a treble player first tackling a tenor to realize and accept the fact that it is an instrument of fewer potentialities than the treble, despite what can be ravishing beauty of tone.

Sopranino recorders

It is more than likely that soon after (or even before) becoming a player of treble, descant, and tenor recorders, you will have succumbed to the charm – and comparative cheapness – of the sopranino recorder in F. It is, however, of limited value. As 'flauto piccolo' it makes a number of delightful appearances in the eighteenth century to accompany the soprano voice, mainly in ornothological contexts, or even, grotesquely, the bass voice (e.g. 'O ruddier than the cherry' from Handel's *Acis and Galatea*). Vivaldi wrote three spectacularly difficult flautino concertos. It appears in Monteverdi's wide-ranging instrumental ensemble for *Orfeo*, and it has a place in medieval music. Modern recorder ensemble music, or arrangements, sometimes require the treble player to switch to sopranino with considerable effect, and immense excitement is similarly produced in the last movement (*Tarantella*) of Gordon Jacob's Suite for recorder and strings (*PB* Ex. 41).

Apart from the possibility of poor tuning a!. ady mentioned, the main technical problem with the sopranino, and even more with the tiny 'garklein' in C above it, is the bunching of the fingers, especially for a player with broad fingers. For an A'-G' trill – 123 456*7, for example, fingers 5 and 7 have to be pulled away from 6, while still covering their holes, in order to allow the trilling finger room to operate. Yet the brilliance of the upper octave calls for display, and therefore for quick and accurate fingering. The tone of the low notes is, of course, thin, and vibrato is needed to give them warmth.

Bass recorders

A bass recorder (more correctly 'basset') in F is needed to complete the normal consort quartet. It is incredible how one bass recorder, sounding comparatively weak on its own, adds a richness of sound, moulds good intonation, and makes the ensemble more coherent, even when there is more than one higher instrument to a part. Given a bass recorder which is reasonably in tune (and many are not), the bass is no more difficult to play than a tenor for the purposes of most consort music. But it does involve reading the bass instead of the treble clef, and the possibility of having to read up an octave if the part goes below bottom F. To play the bass well calls for a great deal of musicianship.

High notes on the bass recorder above D' or E♭' tend to be impure in tone-quality and are likely to need modified fingerings (these are mentioned later in this book). Many basses can be coaxed up to the middle of the third octave (B♭" or B♮") with the use of 'leaking' fingerings.

Renaissance makers cleverly used oblique drilling of the holes to enable the thumb and six fingers to cover their holes directly, using a key only for the bottom note, protected by a massive fontanelle. The key is beautifully shaped rather like a swallow-tail to accommodate both left- and right-handed players. Modern makers have less compunction about using keys, so that perhaps only the middle fingers of each hand directly cover their holes. This eradicates problems of stretch. Generally speaking, the fewer keys the better (but see below). They affect the resonance of the wood, can be noisy in operation, need maintaining, and spoil the simple design of the instrument. Bass recorders, even though made of lighter hardwoods such as pearwood or sycamore, are heavy, and there is a problem of support, usually solved with the use of a sling as well as a thumb-rest. The sling should be adjusted so that the instrument can rest on the right thumb and lower lip without pressure, and so that no support whatsoever is given by the fingers used for playing, or the left thumb.

Basses are made either with a metal crook (or 'bocal'), or with direct blow into the windway which is then sometimes situated at the back of the instrument facing towards the player. Some makers bend the recorder through about 45° ('knick-bass') to avoid undue stretching of the right arm. I strongly advocate a direct blow F-bass, for the distance and directness from the mouth to the tone-creating edge is critical to achieve quick speaking and good tone control. There must be a key for the lowest note, as the hole is out of reach of the little finger, but it should be double to obtain both F and F♯. A further key, or double hole, is desirable to obtain G♯, as the G♯ obtained by half-holing 6 is weak in timbre. These three keys can serve a useful additional purpose for tone and intonation control in the second octave: they may even help to make an upper note speak which otherwise would be unreliable, or else missing altogether. Manipulating these keys can initially cause strain on the little finger, which has to get used to operating in different positions – this is more difficult than using half-holes on the tenor because of the resistance offered by the necessary springing of the keys.

True basses in C, an octave below the tenor and generally called 'great basses', are much more readily available (at a price) than when this book was first published. With an instrument 4½ feet long, a crook cannot be dispensed with, so allowance has to be made for a small delay between tonguing and note-production: the player has to blow fractionally ahead of the beat. With such a large instrument, support is a problem, for a sling, while transferring weight, does not control the swing of the instrument. It is best to fit some form of foot spike so that the recorder rests on the floor (watch that the spike does not slip – a rubber end helps) and then to adopt a sitting position in which either the recorder is vertical, or if not, it leans against the outside of your thigh. The vertical position (between the legs) makes it more difficult to see your music. Either way, the music stand has to be carefully adjusted and placed before you begin to play.

Some makers produce the booming contra-bass in F, an octave below the F-bass, as described by Praetorius, who had a special love for the quiet, low recorders. The contra-bass (which is also referred to as 'sub-contra' or 'gross-bass') is over six feet high, has a long (adjustable) crook, and it is supported on the ground. A great deal of blowing is needed to produce remarkably little noise from such a giant – it seems at first to swallow air from your lungs. The articulation delay is more than the bass in C, but is easily 'thought through'. The effect of these large instruments in a consort is stunning. Yet in the sixteenth century some were fitted with extensions and foot-keys to make them go even lower.

Purchasing a recorder

The following is a concise guide to purchasing a recorder: Christopher Ball's article in *Early Music* (cited in Appendix 2) is more thorough and mentions makes.

If you want to get the best, you will probably have to order an instrument from a world-famous maker with a long waiting list. You have to run the risk that the instrument when it arrives may disappoint you. If both you, and your teacher or experienced advisers, are really convinced that a poor instrument has slipped through, send it back and ask for a replacement. The maker will not want to risk his reputation. Ideally you can arrange with the maker, if he is willing, to visit him when your turn comes up, and, with your adviser, choose from a batch.

Specialist shops in most western countries now maintain excellent selections of hand-finished recorders by a variety of makers. The shop should allow you to try several recorders by different makers, and when you know what you want, then to try several apparently identical recorders by the same maker. If the shop will not let you do this, go somewhere else.

Take with you your own instrument nearest in size, provided it is one you know to be good in pitch and tuning, and an experienced adviser. Having eliminated all but three or four instruments by the same maker test them as follows:

Pitch compare central notes with your own instrument, cold against cold.

Tone play carefully longish notes in the middle of the lower octave, and in the middle of the upper octave, preferably into a corner where the sound reflects back. If you do not like the sound, discard the instrument.

Tuning the instrument should be in tune with itself, and with your own recorder. Check first C to C′, A to A′ and D to D′. If you are satisfied with this test, play the sequence C♯-D♯-E-F♯′-G♯′, especially making sure the last interval is not too wide on normal fingerings. Quite a few recorders are so badly out on this sequence that their imperfections are displayed without further ado.

Speech rapidly repeated staccato C♯'s will reveal if a recorder is slow in speech. With correct tonguing, a recorder should give pure tone instantaneously when breathed into, but give the instrument a chance by using light tonguing. Try the speed of reaction and the strength of low notes such as F and forked fingerings such as B♭. Try a few high notes (E′, F″, G″), or get your adviser to try them, for your experience of thumbing will be less than his. As the instruments you are trying are not warmed up, make allowances for possible clogging on these high notes.

Weak notes and 'Wolf-notes' examine the lower octave again (all notes chromatically) for evenness of quality. A note may be weak beyond redemption (?B), or some notes may be unstable with an inbuilt vibrato of their own (?G). Bass instruments are more liable to such faults than trebles.

Volume and flexibility make crescendos on F′, C, and C′ (others if you like) to see how loud you can go before the note breaks, and in the first part of the crescendo how much pitch changes. Judge how resistant the voicing is. Is it what you want?

Alternative fingerings see that alternative G', E, and D are usable as regards intonation and tone-quality.

Construction check that the wood is of close parallel grain with no knots. Look for incipient cracks, and see there are no loose splinters of wood round the windway, edge, and holes. Joints should be tight, smooth, and snug, the plug not loose to the touch, and the instrument well balanced as you feel it in your hands and under your fingers.

This check-list may seem long, but following it systematically should not take more than three or four minutes per instrument, and will help you to leave the shop well satisfied, even if you place the manager in a moral dilemma over the recorders you have discarded.

Care of the instrument

Good technique can never overcome the faults a recorder will develop if it is not properly looked after. Pages 1–6 of Edward L. Kottick's *Tone and Intonation of the Recorder* make the same point and deal thoroughly with recorder maintenance. This section covers the main points in less detail.

First and foremost, keep your recorder in a stout wooden box that shuts firmly. Any container that does not protect the recorder when the box is dropped is useless. Never leave a recorder on the floor or on a chair or perched on a music stand. Do not let a recorder get hot: heat dries out natural wood and makes a recorder made of impregnated wood sweat drops of paraffin wax.

After playing, *always* wipe the recorder bore dry with a mop or piece of material that does not shed fluff. Do not force a mop into a recorder nor push it into the head-piece up against the plug. A good recorder wiper is a linen handkerchief rolled diagonally and twisted gently through the bore of the instrument or put over a small mop: it leaves no bits, dries moisture, and, above all, polishes the surface of the bore. Never use the slightest pressure or force, as this could deform the shape of the bore. A square of chamois on a weighted string is also an effective bore dryer, cleaner, and polisher. A sopranino made in one piece may have to be wiped dry with a pipe-cleaner, possibly doubled on itself; take great care to watch through the window and stop short of the plug. Bass recorder crooks should be shaken out, and left on their own to dry.

To keep the windway free of fluff and deposit, it should be wiped out with a small, soft feather, but not one with so thick a spine that it touches the floor or roof of the windway. Take care not to let the feather push against the edge. Accumulated deposit at the corners of the windway and the edge should be eased off with the quill of the feather, and with infinite care. After each playing, always wipe out the windway with a dry feather. Leave the recorder in pieces to dry out. The day before a performance, or other playing session, a feather dipped in a weak detergent solution should be passed through the windway to remove patches of grease which attract moisture, and immediately afterwards the windway should be washed through thoroughly with a little warm water so that no detergent whatsoever remains (it might damage the wood): shake out as much water as possible from the windway and bore, wipe the outside of the head-piece, use a dry feather in the windway, and allow the head-piece to dry out completely. It is very important that this process should be carried out regularly, and at first frequently, with a narrow-windway baroque recorder. This, combined with high wind-pressure in playing, should prevent clogging. Never 'blow out' a clogged recorder for this only spreads moisture; use a feather instead, or, if it clogs while playing, suck sharply in. Similarly, warming up a recorder by breathing into it only defeats the object of dry warmth before playing.

Oil the bore of a recorder, other than one made of impregnated wood, when it is just played in and about twice a year thereafter – more frequently when the wood tends to soak up oil, less frequently with older recorders. Use a non-softening, non-acid, resinating plant oil such as banana, almond, or unboiled linseed (not olive): these oils leave a thin protecting film on the surface of the bore, helping to keep it smooth and polished. If too much oil is used it dries sticky and attracts fluff, as well as deadening the benign influence of the wood-spirits. As the outside of a recorder is usually varnished, oil cannot affect it either beneficially or adversely, although an old recorder with the varnish worn off does benefit from occasional oiling, and a new one looks smarter. An unvarnished recorder, in particular its head-section, needs to be well nurtured with oiling, especially when new. *On no account* let oil, or even your fingers, touch the tone-producing regions which must remain as pervious to moisture as possible: oil on the edge and thereabouts encourages the formation of globules, those recorder player's 'gremlins'. It is best, too, to keep oil away from the plug which might be over-loosened by lubrication. When applying oil, warm it in the palm of the hand first so that it becomes thinner and penetrates the surface of the wood before drying. The recorder also should be absolutely dry and warm. Oil should be applied to the bore with a mop used only for that purpose.

Joints should be airtight, and should be snug enough not to allow 'play' in the lower sections of the instrument, particularly the foot. Never hold the instrument by the foot joint, which weakens it. Instruments should both be assembled and taken to pieces by screwing gently but firmly in the direction of the lapping of the joint (clockwise), without any sideways force. If cork joints are loose, they may be made to swell slightly with moistening, or can be burred with a pin: another method is to grease the cork and then warm it lightly over a match flame. Thread joints can be varied at will by winding off or adding thread; bobbins of soft and resilient waxed thread are obtainable from woodwind suppliers, or else cobbler's hemp rubbed with beeswax or soaked in melted candle wax may be used. If the inner section of a joint develops a hair-line crack, it should, as a temporary measure, be bound tightly between the lapping and the bottom of the joint with fine and inelastic wire such as five-amp fuse wire. A hair-line crack on the outer sleeve of a joint, or any larger crack, is more disturbing and the instrument should at once be returned to the maker to be repaired (e.g. to be fitted with a retaining ring) or replaced. If a cork joint becomes so worn that a section of the instrument is loose, this too is a case for hospital treatment at the maker's, not just first aid with pieces of thin paper. Cork joints can be preserved by applying lanolin, enough of which can be obtained inexpensively from a chemist to last the recorder's lifetime. Lanolin also makes a stiff cork joint easier to operate and it is a good idea to put some on the joints of brand-new instruments before assembling them. Soap or vaseline have the same effect.

After a certain amount of use the thumb-hole of a wooden instrument wears away with the rubbing of the thumb-nail on it. Experienced players do not 'thumb' violently so they are less troubled with this disorder. With a badly worn thumb-hole it is difficult to judge the width of the space left between the thumb-nail and the edge of the instrument ('thumbing aperture') and high notes are therefore hard to form. Tone may be affected if the thumb-nail intrudes too deeply because of a worn hole. The instrument maker should then be asked to 're-bush' the thumb-hole, preferably with a hard ivory substitute.

Keywork will need occasional oiling with a sewing machine oil. Cork stops need a spot of lanolin. At long intervals leather pads will need replacing by an instrument repairer.

New wooden recorders need playing in, to break the wood gently to its task of maintaining its personality under varying conditions of temperature and humidity. However tempting it might be to go on, a player with a new instrument should stop playing after fifteen minutes for the first week or so of the recorder's life and dry the instrument thoroughly (swab and feathers) before putting it away unassembled. A new recorder should be played every day, increasing to half an hour in the second week. A month-old recorder properly played in will last out a normal playing session without danger of the wood cracking. With care, however, its tone will continue to improve as apparently it adapts itself to its owner and as the owner comes to recognize its qualities and idiosyncrasies.

Playing position

Playing position is described and illustrated in *Introduction to the Recorder*. But to recapitulate and develop:

Be relaxed in your body and arms, right down to the fingers and especially the thumb.

Never raise your shoulders, or become hunched.

Play with the recorder at 45° or less to the vertical, with the elbows beside but slightly away from the body.

Wrists should be low (think of your hands holding a soft ball of wool).

Finger tops should be near horizontal, and fingers almost at

right-angles to the instrument (as nearly as feels comfortable). All finger-movements in playing should be short in
distance, quick, light, and economical.

Be seated comfortably in a firm chair without arms. Do not
lean right back while playing: be slightly forwards to your
music – or rather the audience, for the music stand should not
come between you and them, dampening sound projection.
But do not play the recorder from the side of the mouth, as
this may impede the passage of air, and looks wrong. If you
stand, adopt a stance in which you are balanced between
both feet, one slightly forward and at a slight angle to the
other. Standing helps breathing, but if others are sitting, it
can imply dominance. Consequently, you should stand for a
concerto, but sit for ensemble music (as a string quartet
does). Remember the baroque solo sonata is usually
conceived as ensemble music with the bass equal in
importance to the treble: if you prefer to stand, don't feel you
are playing a concerto.

The recorder is supported by the right thumb and lower lip,
the latter in a relaxed, slightly dropped position (see later).
Although the little finger of the left hand occasionally gives
lateral support, the operating fingers and left thumb should
have nothing to do with supporting the recorder: all their
attention is taken up by playing, and their movements must be
unimpeded by other considerations. Basses need slings, great
basses slings and floor contact; but all recorders, with the
possible exception of the descant and sopranino, need a
thumb-rest.

Placing the thumb-rest depends on where each player feels it
gives him the most relaxed support. This is usually under and
between holes 4 and 5. The thumb-rest may be slightly at an
angle, or even slightly off-centre. Before fixing it permanently,
try it in various positions held temporarily with Sellotape or an
elastic band. It is a common fault to put it too high in the first
place. This allows the right-hand fingers to be held too
obliquely to the recorder for efficient movement. Thumb-rests
can be of cork, or cork-faced: they should not be metallic or hard
where the thumb touches them. They have to be broad enough
to give comfortable support in long use. Clarinet-type thumb-
rests can be bought and fixed with contact glue, not screws; or a

cork can be fashioned and taped on (Kottick, p. 9). With a descant, the surface tension of the thumb may give just sufficient support if the instrument is light, or a piece of Blue-Tack may be moulded into a thumb-rest shape and taken off after playing. Trebles and larger instruments ought to have fixed thumb-rests, despite their ugliness, for fine playing. It is a pity that makers do not supply a shaped piece of matching wood with cork facing for the owner of the instrument to glue on when he is certain where it best suits him. Moving a thumb-rest previously fixed inevitably makes an unsightly mark.

Before playing, ensure that the holes are in line with the centre of the windway. This looks best, and is probably the most comfortable position for your fingers: it will not affect tone if the holes are slightly out of line with the windway, but it will certainly do so if you do not properly cover a hole with the pad of the finger, or if a key is leaking. If you suspect the latter, take the head-joint off, cover tightly the bottom end of the bore (e.g. with your bare knee) and blow into the middle section, all fingers on as if for bottom F. No air should emerge. If it does, shuffle your fingers and press each key in turn very hard to identify the fault. A key may need repadding.

The foot section should be carefully turned so that it best suits your little finger both for F and F♯. It is worth while making a small mark on the middle section and on the foot section at the join so that, once found, alignment is always correct. A slight move either way will result in your not obtaining the bottom notes accurately.

During playing there is no need for the player to move much, except if he is leading, in which case he needs to convey to his ensemble, by small but clear movements, the start and finish of the music, time changes, and so on. If he is relaxed and lost in the music, he will not be completely still, but will show his involvement by some spontaneous, almost unconscious move-ment. This should be above the waist, and should on no account penetrate to the feet, which may consciously have to be prevented from beating time. The pulse of the music, established in the second before starting, is an internal feeling, though shared by all the players in an ensemble. This realization that a piece of music is a 'going concern' among all those involved (including the audience) is in itself very rewarding to the player.

II

BREATHING

Two factors affect how one breathes in order to play a wind instrument—first the expenditure of breath needed to make a note, and secondly the pressure required. In oboe playing breathing is a major problem as the instrument uses a small amount of air at a high pressure, and this is true to varying degrees with most other wind instruments. Recorders, however, 'go with a gentle breath', using as much air and at as low a pressure as is needed for reading aloud or soft singing. Were it not for the fact that intake of air must be very much more rapid than exhalation, it could be said that in playing the recorder one breathes naturally. Recorder playing is just unnatural enough to cause breathlessness or indigestion if it is indulged in immediately after eating or drinking.

Breathing for recorder playing, as for singing and playing other wind instruments, should be from the bottom of the lungs, that is, the diaphragm. Sit or stand upright with the shoulders back and the head up, and become conscious of the movement of the muscle across the triangle formed by the ribs below the breastbone. Breathe in deeply through the nose so that the diaphragm is fully extended but do not raise or hunch the shoulders in the process: then, with your hand on the diaphragm, release the air in five or six separate and fairly slow exhalations, noting the muscular movement. Now open the mouth and throat, and, using the diaphragm only, draw in a good breath as quickly as possible so that the lungs are nearly full. Half close the lips and let the air go out softly, slowly, and evenly. Notice that in breathing in both directions the muscles of the throat and mouth do nothing: they remain relaxed so that air can be pumped past them by the diaphragm. It is easy to fall into the bad habit of snatching a small amount of breath in quickly with the throat – that is, gasping – when a breath has to be taken during a very short interval in the music: it is vital, however, that even when time allows only a little replenishment of the lungs the action should come from the diaphragm.

The impression one should have when blowing into the recorder is that the breath originates from the base of the lungs, a sensation of air forced up from underneath and travelling smoothly up the windpipe, through the throat, across the mouth and without hindrance into the windway of the recorder. The conscious realization of this sensation of the passage of the air from the diaphragm to the edge helps to keep a long note steady. If you have difficulty in experiencing the sensation, artificially create conditions in which the diaphragm has to pump harder by making a stricture in your mouth with your tongue against your teeth, and feel the push of the air up to the stricture, and through to the instrument. As an aid towards playing a long note evenly, this device would be more useful if it did not disturb the airflow and create unwanted eddies that affect tone (see also pp. 103 and 104); it can serve a purpose, however, where very quiet long notes are required, as diaphragm control at ultra-low pressures is extremely difficult and in such circumstances good tone must be sacrificed for steady intonation. Some further assistance in keeping a long note steady can be gained by imagining that the note is being sung, quietly, to the sound 'aah'. Once a note is started, the tongue should return to this 'aah' position, in which the tongue and the lips are well placed for good tone production, at least of low notes (see below).

When playing a piece of music, a recorder-player should feel all the time that his lungs are reasonably full, though never bursting. Subject to constraints of phrasing, no opportunity where breath might be replenished should go by unused. Most rests are breath marks. Aim at breathing frequently even if only a little at a time, so that occasions where a big (and noisy) inhalation takes place are rare. It is particularly important to breathe often and deeply when playing bass instruments: breathlessness causes anxiety both to performers and audience. With a sopranino the condition might arise where a lungful of air becomes stale before it is used up, and like an oboist, the player has to breathe both out and in at the end of a long phrase.

Breath-pressure

While many players tend to blow too hard, others (perhaps to avoid disturbing their neighbours) develop the bad habit of blowing too softly in relation to their instrument's best tone-quality. It is vital to establish the optimum breath-pressure for each note, and an optimum overall breath-pressure for your instrument's *mp/mf*. Practise as follows: finger the note G′, and, after breathing in deeply, whisper into the instrument to make the quietest noise you can imagine that passes for a musical note. Hold it *steady* (this is very difficult) for twenty seconds; don't let it get out of control at the end. Now play the same note loud like a trumpet, giving it the greatest breath-pressure you dare, almost making the note 'break' – that is overblow with a nasty rasp into a higher octave. Hold the note steady for ten seconds (this is relatively easy) and finish by blowing it over the break. The first note will be thin and wispy, the second strident, and the two should be at least a whole tone apart in pitch. Now go back to the low breath-pressure but make the note just loud enough to be pleasant and convincing (it will still, however, be a little breathy): this is your instrument's *pp*. A clear, loud (but still slightly harsh) note well clear of that break you have just discovered, is your *ff*. Somewhere between the two pressures, generally but not necessarily midway, lies the most beautiful G′ – find it by experiment – and this is your optimum breath-pressure. In consort playing, however, your general level of attack should be slightly below this pressure, so that your purest notes come in an *mf* passage or when you announce a theme in a Fancy.

When you have found your best G′, move on to other notes on the instrument, studying their behaviour under varying breath-pressures. You will thereby discover the level of pressure at which your recorder plays best. As stated in Chapter I, different recorders are voiced to play best at different breath-pressures, so be prepared to vary the general level of attack to suit each instrument you possess.

To consolidate these experiments play at low, high, and medium pressures and at a speed of about five seconds to each note any slow hymn-tune made up of equal length notes (e.g. 'Rock of Ages' starting on C, or see *PB* Ex. 1). Practise on all the recorders you have, and concentrate on steady breathing at low pressures, as this is more difficult to achieve. Play towards the corner of a room, so that the sound is reflected back at you; concentrate very hard on listening to it (would that more musicians did this!) and be extremely critical. Do not allow a waver in the evenness of your sound. Try to be aware of the undertones and harmonics that form a constituent part of each note. Hearing these and keeping them steady and in tune may help you in achieving your best, non-vibrato tone-quality for each note. Include in this process all the normal-fingered notes up to G″ and the common alternatives, E, G′, and D.

Having established the optimum breath-pressure for each note, practise the rapid change in breath-pressure required when slurring from a note in the lower octave to a note in the upper octave above the 'break' in registers, e.g. F′ to A′. To avoid 'clicks' two conditions are essential – extreme rapidity of finger movement and a sudden though small increase in breath-pressure from the optimum for F′ to the optimum for A′. Some recorders behave better than others with awkward slurs over the register break: a player who can do these slurs on any instrument pianissimo without 'clicks' may claim to be an expert.

Another breath-pressure technique that must be mastered is the tiny decrescendo that finishes the last note of a phrase: it must be so miniature that a listener does not get the impression of the note going flat. A momentary holding-back of breath-pressure is useful in negotiating difficult slurs, for, if breath-pressure is reduced for that fraction of a second when the fingers are moving, 'clicks' are made less prominent if they do occur. Practise wide slurs such as F′ to B♭ or D′ to F′ and notice how 'in-between notes' can be eliminated by breath-pressure control. This problem of crossing register-breaks will be referred to again under 'Tonguing'.

Vibrato

Recorder-players who, anxious to obtain the best and most characteristic tone-quality their instrument will produce, sensibly seek to improve their breath-control by reading books on vocal technique (e.g. *The Singer and the Voice* by Arnold Rose), will discover that some teachers list vibrato as a defect caused by nervous debility. This harsh judgement probably stems from the fact that vibrato can develop into a bad habit if it is used without moderation and discretion. To a recorder-player, vibrato is as important as it is to a string player. Its use is discussed in Chapter VII: its production is described here.

Vibrato is normally produced by alternately decreasing and

increasing breath-pressure, regularly and rapidly.[1] Although vibrato can be effected by tongue movements, or by finger movements (the 'close shake' as advocated by Hotteterre), the most usual modern way of producing it is from the back of the throat, i.e. from the larynx. The action may be represented by the syllables 'hu-hu-hu-hu-hu' aspirated deep in the throat without any interruption in the actual flow of air. The effect is easiest obtained on a sopranino or descant at high breath-pressure. Start with a 'long wave-length' (slow 'hu-hu's') and increase the frequency until you get a 'short wave-length' of vibrations – this is the natural vibrato that some beginners produce without realizing it (they find it all the harder to play a plain note without vibrato). Once the knack is acquired (and it is deceptively easy) you will soon acquire a comfortable short wave-length vibrato, and you should be able to produce notes of a warmer tone-quality than you did without vibrato. Never let vibrato become entirely automatic. Continue to practise long notes without it (they are much harder to control without vibrato just as it is more difficult to balance on a bicycle without slight wobbling): playing without vibrato is essential in medieval music, in most consort playing, and frequently in baroque music to achieve expressiveness by contrast of tone-quality between 'warm' and 'cool' notes in a phrase.[2] Practise also the other extreme – slow vibrato (long wave-length): aim at keeping the vibrations steady and even, at about six to a second without getting any quicker, continuing to use a high breath-pressure to aid control. Attempt, too, something between this slow vibrato and more normal rapid vibrato (call this 'medium wave-length'). The following diagram may clarify matters; note that the 'long wave-length' vibrato is generally

---

[1]There is a fuller discussion of breath-control and vibrato on pp. 46–57 of Daniel Waitzman's *The Art of Playing the Recorder*.
[2]Since I first wrote this book I find I use less vibrato than I used to, recognizing that many eighteenth-century writers regarded vibrato (finger-vibrato more than throat vibrato) as an ornament for gracing long notes and the climax of phrases. But historical evidence – as so often – is conflicting on this subject. It is, however, salutary to remember that Leopold Mozart would have expected his son to be able to play the violin without vibrato. The ideal to aim at is a beautiful note without vibrato. This can then, where it is musically desirable and historically proper to do so, be accorded the additional warmth, intensity, and variety of vibrato.

played with wider fluctuations than the 'short wave-length'
vibrato:

long wave-length: ⌒⌒⌒⌒⌒⌒

medium wave-length: ⋀⋀⋀⋀⋀⋀⋀⋀

short wave-length: ⋀⋀⋀⋀⋀⋀⋀⋀⋀⋀⋀⋀

no vibrato: ━━━━━━━━━

One can, however, control the amplitude of vibrato (i.e. how far above and below the central pitch of the note the vibrato ranges) as well as its wave-length. In general, aim at equal fluctuations above and below the central pitch, although in some musical contexts vibrato may be more effectively applied (by alternating reductions of breath-pressure) to the flat side of the note, as happens with finger-vibrato. You should master the art of moving with gentle gradation from one character of vibrato to another. Start this by doing eight vibrato pulsations to a minim (semiquavers), then sixteen (demisemiquavers). Then try twelve (in triplets). Finally, control the pulsations as a regular acceleration from long to short wave-length, merging off into a plain note. Then do the exercise in reverse: try the effect of beginning a note without vibrato, then introducing into its cool timbre a small and very rapid vibrato which gradually slows down and widens into a full-bodied long-wave vibrato; this exercise is easier if the note swells in volume at the same time. Apply the three vibrato wave-lengths to your hymn-tune practice, playing at the three different breath-pressures. Counting the plain note, this gives twelve ways of playing the same hymn-tune: listening very minutely to what you are doing, try them all. If you work hard, for the process of mastering all twelve interpretations is a gruelling one, you will find that the simple recorder becomes in your hands a truly expressive instrument.

III

TONGUING

The last chapter described only the process of making the middle part of a note. This deals with the beginning and ending of a note on a recorder—how to 'give it breath with your mouth', and to take it away.

The consonant to aim at for normal tonguing is a whispered 'dh', obtained by placing the tongue in the position to say 'd' but then at the last moment changing it to 'h'. Try this with the second syllable of the word 'London' in mind, quietly whispered to yourself, and with the 'er' vowel sound held on for a second or more. It is a very soft sound, almost imperceptibly a consonant. Yet, within limits, it can vary in strength (compare the 'dh' sounds made in 'dhaa', 'dher', and 'dhee').

Strong tonguing is easy to do – in fact those very words 'to do' are instances of it. Both 'd' and 't', like 'dh', may be played with different degrees of 'explosion', so extending the usual range of tonguings. Light tonguing, softer than 'dh', is harder to master, but its acquisition is imperative for playing the lowest notes of the recorder and any notes requiring cross-fingerings ('forked' or 'gapped' fingerings), e.g. low Bb, Eb, high C♯', etc. The more fingers that are below the hole left open the lighter the tonguing required. Try, for example, the notes produced by the fingering 0 –23 4567 arriving at it by descending from E 0 –23 –––– and adding one finger at a time; with the faintest tonguing imaginable, the note should be somewhere near Bb, while a slightly more definite tonguing should give F♯'. Observe that the note produced seems to depend more on the tonguing than on the breath-pressure used. Now, taking off a half-hole at a time and using the very light tonguing, play a series of notes that will be in the vicinity of B♮, C, C♯, and D. Once struck, play them as loudly as you can without causing them to break upwards. In the slightly stronger tonguing you should get a similar range of less than semitones from F♯' to A': once struck, play them softly. This illustrates the important principle that strength of tonguing in articulating a note does not necessarily vary with the breath-pressure accorded to the note itself. Practise jumping from one 'register' to the other in quick succession over these five pairs of notes: do this so that each note speaks clearly and immediately.

To appreciate gradations of tonguing from light to strong try the following exercise. Starting with a good lungful of air, play, without vibrato, the note G', keeping the tongue still. Gradually move the tongue backward and forward without quite letting it touch the protruding ridge in the gum (or front palate) above the front teeth: this introduces 'tongue-vibrato' into the note (see p. 107) and may be termed 'y' tonguing. Now bring the tongue movements forward so that the tip of the tongue just grazes the very angle of the teeth-ridge, having the effect of impeding but note quite stopping the flow of breath: this is the whispered 'r' position of tonguing, the lightest possible tonguing (for 'y' is a sort of non-tonguing). The 'r' is not the English dental fricative nor the French glottal 'r', but more like the Scottish 'r' as in 'baron'. By gently pressing the tongue slightly forward at the same point of contact on the teeth-ridge you will come into the 'l' tonguing position in which the initial flow of air is actually stopped centrally but is only impeded at the sides of the tongue. Next go to the whispered 'dh' position, which may be regarded as normal single tonguing. The 'dh' position is the same as for 'r', except that instead of grazing against the teeth-ridge and impeding the airflow the tongue touches the teeth-ridge and just (but only just) cuts off the gentle flow of air momentarily, with the greatest possible delicacy. 'd' itself is more deliberate, being a sufficiently firm contact of the tongue upon the teeth-ridge to build up a slight pressure of air immediately before the release of the tongue to enunciate the note. In the 't' position the tongue is more rounded, less flat, and thus touches at its tip between the teeth-ridge and the top of the front teeth. Note that in no tonguing position does the tongue actually touch the teeth. 't' allows for a greater pressure of the tongue and consequently a greater build-up of air pressure before the moment of release, i.e. it is in character the most plosive of tonguings. Yet, like all tonguings, it has a wide range from a Gallic spit to its softness in a quietly-whispered 'butter'.

Practise this range of graduated teeth-ridge tonguings, in both directions, on a repeated G', attempting to maintain constant volume. It requires determination to avoid a crescendo or decrescendo, but try it at different breath-pressures, i.e. a constant *mf* throughout the exercise, then a constant *f*, then a constant *p*. If 'aa' is used to indicate the flow of breath after tonguing, the exercise may be expressed thus (forwards or backwards): 'taa-daa-dhaa-laa-raa-yaa-aa', each syllable except the last being repeated three or four times as part of a gradation throughout.

It would be a convenience in recorder playing if one's mouth were differently designed so that this exercise could be accomplished with very short strokes of the tongue. But the tongue has to travel in order to get from grazing the teeth-ridge to the 'aa' position where it least impedes the airflow, that is to say, at the bottom of the mouth with the tip touching or near to the soft palate under the lower gums. It is quite a long journey from here to the upper teeth-ridge; but it should be made very rapidly and slowed only on the point of graze or contact (like a car braking suddenly to avoid a collision). As all doctors know, the 'aa' position gives the largest mouth cavity with the jaw and cheeks relaxed, and the most cavernous opening for air to travel through on its way to the recorder windway. As we shall see under 'Tone', this seems to favour low notes, giving them their richest quality. When you play high notes, or when you play softly, you will, in imitation of a singer and to maintain control of breath, use a less hollow and a more shaped and a narrow wind-passage through the mouth. This enables you to use a smaller traverse in the tongue stroke. See where your tongue is placed in saying 'dher' and in saying 'dhee'. With 'dher' it is higher in the mouth, nearly touching the lower part of the lower teeth. With 'dhee' it is central in the mouth, nearly touching the top of the lower teeth. From the nearly touching 'ee' position, it is only a small stroke to the upper teeth-ridge, facilitating rapid playing in the upper and middle notes of the recorder, where most passage-work fortunately occurs.

The position of the lips is related to the tongue positions, to the shape of the beak of the recorder, and to the necessity of opening the mouth wide for the diaphragm to draw in quick deep breaths. In earlier editions of this book I used 'oo' as the mouthing vowel after a tonguing: I have now chosen 'aa'. No vowel in fact is quite right in relation to the lips. 'oo' is more forward in trajectory (and the tongue is low in the mouth, favouring low note tone), but one does not round one's lips in putting the recorder to the mouth. 'oo' brings the cheeks in, whereas they should normally be slack, as should the lower jaw, in readiness for opening far enough to draw in a quick deep breath. 'aa' indicates this labial relaxation, but the lips are too wide apart to surround the mouthpiece. 'er' better indicates normal lip position on the recorder, but gives too little oral cavity for low notes and is too slack a sound to encourage firm, outgoing playing.

The lips should be slack enough to be moulded into shape by the recorder's mouthpiece, but then firm enough to make the contact absolutely airtight. The beak of the instrument should be just far enough into the mouth to make it impossible for the lips to close behind the windway, i.e. to enunciate 'p', without forcing the recorder foward. If it goes in further, the lips will not be able to form themselves gently as if they were the same shape as the windway opening. Do not seem to be swallowing the mouthpiece. Except for certain tone productions and very quiet playing (see p. 104), the teeth should be apart so as in no way to affect the flow of the column of air from the lungs to the windway. If the teeth impinge upon this flow of air they may form eddies which could affect tone-quality. Their distance apart may narrow slightly from 'aa' to 'ee', but not so as to disturb the seal of the lower lip upon the mouthpiece; jaw movement is rather similar to chewing with the lips closed. Latitudinally there is movement at the ends of the lips between 'aa' and 'ee' as the lips widen. This makes the lips slightly tighter on the mouthpiece for high notes, slightly more forward for low notes. A recorder player seems to smile at high notes, and become more serious for mellow low notes.

If these lip positions are kept firmly in mind, and the lips, as it were, held set in one of them, it is possible to practise tonguings without the recorder. You can thus practise one of the most important aspects of recorder technique at any time (in private). Keep the lips still, and whisper – do not say or voice – your tonguing syllables or mnemonics. Use exactly the same flow of breath as you would in actual recorder playing.

Certain notes on the recorder require very careful tonguing. One of these is C♯′, or, even more pertinently, G♯′ on the tenor. This is a peculiarly slow-speaking note and unless it is tongued very gently will either cough before speaking or will strike a tone and a half too high. Passages where this note has to be repeated rapidly need cautious treatment, and it is wise not to break the flow of air after the first tonguing, saying, as it were, 'dhoo-yoo-yoo-yoo' instead of 'dhoo-dhoo-dhoo-dhoo'. The same is true of repeated high F″s. Special tonguings are needed for the highest notes of the recorder, including 'whoot' for the very high C″ (see Chapter VI). A trick to soften tonguing, particularly when breath-pressures are high (e.g. in a series of high notes), is to breathe out momentarily through the nose at the same time as tonguing. This is a useful anti-panic device when a high F″ is looming up, though the solution to this note is relaxed thumbing.

A player should also be able to start a note without tonguing, aspirating 'h' from the throat. This technique has a place in quiet playing on low instruments.

Attack and rhythm: portamento and staccato

Any musical note may be thought of as having extremes of attack, length, and loudness. Between each of these extremes (e.g. volume from *ppp* to *fff*) lies a wide vocabulary of different ways of treating a note, each suited to a particular context. Tonguing on a recorder constitutes articulation or attack; it is equivalent to bowing on a stringed instrument. Tonguing combines with variation in the length and volume of notes to create rhythm. In a bar of four crotchets in common time all on the same note, rhythm is primarily established by the degree of attack accorded to each note. The attack on the first note is more definite, and it is played longer and louder than any of the remaining three: a secondary emphasis is placed on the third note, while the last is the slackest, shortest, and quietest of the four. On the note G′ practise bars containing one note-value in 4/4 time, 3/4, 6/8, and slow 2/4 ('One – and – two – and'), and by using variations in tonguing, length of note, and volume, establish rhythms. Try this at different speeds, and in different styles between legato and staccato, from *ff* to *pp*. Then attempt to do the same thing with tonguing only, using notes of equal length and loudness. If, *using tonguing only* and playing only the note G′, you can communicate to a friend the differences between a 4/4 bar and a 2/4 bar at the same speed in quavers, or between a 6/8 bar and two 3/4 bars at the same speed, you have

acquired considerable subtlety in tonguing control. Gradation
of attack such as that which differentiates

by means only of tonguing, illustrates nuances of technique in
the service of interpretation (*PB* Exx. 9 and 10).

Tonguing is not only the main ingredient in establishing
rhythm, but it also controls rhythmic subtleties such as playing
a note slightly before or after the established beat rather than
exactly on the beat. Used with discretion, this device can, as
every conductor knows, increase the expressive impact of music.
A note in a slow movement (e.g. a Siciliano), fractionally
delayed in its articulation, can cause a *frisson* of pleasure to the
listener: if it happens too often, it becomes a nagging
mannerism. A note played fractionally before its beat can, if
used with discretion, give an impetus and excitement which
mechanical slavery to the beat would never create.

Long notes barely separated by light tonguing constitute
'portamento', the word deriving from the fact that one note is all
but carried over, or slurred, to the next. The musical indication
is

although portamento tonguing is used for playing any legato
passage: in one with a series of notes of equal written length,
such as the example quoted, the effect should be such that the
'dhaa' tonguing almost becomes 'lhaa'. The tongue itself should
be made to feel soft and flabby in portamento tonguing, only
just grazing the roof of the mouth for each note. Fingering must
be quick or the notes will trip over each other. Portamento is an
important technique to acquire not only for its interpretative
applications (*PB* Exx. 11 and 12) but also because the ability to
tongue with an infinitesimal interruption of sound is invaluable
in suggesting a slur over notes that would (for lack of alternative
fingering) be extremely difficult to slur properly without

'clicks'. An instance where this form of deceit might be used is the following bar from No. 8 of 'Fifteen Solos' (Schott):

Only the most brilliant players could play this slur perfectly without any suggestion of sound other than its three notes, and without stressing the top D'. With a touch of tonguing between the notes the player can eliminate 'clicks' and deceive most listeners into hearing a slur, particularly if he plays the remaining three notes of the phrase fairly staccato.

Staccato may be considered the converse of portamento, since it consists of short notes played with strong to medium tonguing. One of its developments is 'echo tonguing'. This is the recorder's equivalent to the harp pedal on the harpsichord and is used for echo effects in passage work. Echo tonguing is produced by pressing the tongue firmly on and below the teeth-ridge (in the 't' position) and releasing it momentarily and only enough to allow a little air to pass to produce a short and stifled note. The impression to the player is that his tongue is almost drowning the sound of the instrument by the noise of its activity.

Tonguing in terms of duration of articulation

Up to this point we have considered tonguing in terms of 'strong' and 'light'. We have characterized 't' as a strong tonguing and 'r' as a light tonguing. But we have noted that each tonguing is susceptible to a wide range of strengths depending on tongue pressure and the extent of build-up of airflow immediately before the tongue is drawn back to articulate the unimpeded flow of breath which makes the body of the note. Thus in whispering 'London' the stressed first syllable requires the 'l' consonant to be stronger than 'd'. Nevertheless 'd' is inherently more plosive than 'l'. Put another way, however hard you try, you can never make 'l' so strong that it is stronger than the strongest 'd', nor make 'd' so light that it is lighter than the lightest 'l'. Thus at each extreme of tonguing there is a forcefully spat 't' and a barely enunciated 'y'.

But for a full understanding of the art of tonguing one needs to consider another dimension. This is in terms of how long (in microseconds) it takes to say the tonguing. In these terms 't' may be described as sudden, precise, sharp, quickly articulated. 't' is always precise and quick-speaking, whether it is said strongly or lightly. 'd', because in normal speech it is voiced and a little space of time is needed for the larynx to give voice to it, takes longer to articulate than 't', even when it is whispered. 'l' and 'r', both voiced in normal speech, take longer to enunciate than 'd'. They are slow-speaking tonguings, whatever their strength or lightness. 'y' is so slow to articulate that its status as a consonant is at issue: it is a semi-vowel. The slow-speaking tonguings are less precise, making them difficult to use in upper-register selection for the neat articulation of high notes. But, as Hotteterre points out, they imbue a note (especially a middle or lower note) with a softer tone.

We must now look at other tonguings used as components of double-tonguing, that is to say, the alternation of two tonguing consonants, the second taken on the rebound, as it were, from the first. These were sometimes referred to as 'direct' and 'reverse' tonguings. Thus 'd' or 't' could be alternated with 'r', i.e. a quick-speaking tonging may provide the impetus for a slow-speaking tonguing on the rebound. For passages of gently undulating smoothness, 'l' was the direct tonguing followed by 'r', both slow-speaking tonguings. Writers from the sixteenth to the eighteenth century exhorted recorder-players to practise their 'diri liri's', for double-tonguing with 'r' was, and perhaps should still be, even more important in recorder playing than single-tonguing. Even nineteenth-century flute tutors advocated 'territory' as a mnemonic.

Reverse tonguing is an especially appropriate description of the palatal tonguings 'k', 'g', and 'gh', which are produced by the tongue against the back of the roof of the hard palate in the position it reaches at the end of the releasing or withdrawing stroke from a teeth-ridge tonguing. They alternate with 't', 'd', and 'dh' respectively. This alternation between a forward and backward tonguing helps the player to feel the rebound of double-tonguing, and provides a tonguing differentiation sufficiently marked to achieve a greater feeling of control over fast semiquavers than can ever result from single-tonguing. Yet in terms of duration of articulation, 'k', 'g', and 'gh' are exact counterparts of 't', 'd', and 'dh'. They are slightly less plosive, but are each as quick-speaking as their partners. 'k' is the most precise; 'g' is slightly less so because it is voiced in normal speech; and the aspirated and less plosive 'gh' is only a more whispered 'g'. They enable groups or passages of short notes to be played at a higher velocity of double-tonguing than is physically possible with single-tonguing. In the twentieth century, when the prime function of double-tonguing as a means of achieving control and expressiveness was largely lost sight of, 't-k' and 'd-g' pairings became the only form of double-tonguing in their pyrotechnical role.

To complete this parade of tonguing consonants (ignoring Ganassi's 'head-breath', or 'p' tonguing), there is the eighteenth-century 'dl', a slow-speaking reverse tonguing paired with 'd', 't', or 'l' as in 'tiddle-diddle-liddle-diddle'. It is similar to 'l' in that airflow is released at each side of the tongue, but it must be linked with 't', 'd', or 'l' for its completion: it can therefore only be used for groups of equal-value notes at a fairly quick speed. Because it is a reverse stroke *par excellence*, it will, when mastered, smooth up passage-work in a highly gratifying way, but as it is slow-speaking it cannot cope with passages of great velocity, nor with high notes. 't', 'dl', and 'l' together make an easy-swinging triple-tonguing, useful in jigs, pronounced as a three-syllable 'tiddly' or 'diddly'.

It must be stated that it is possible to play the recorder well by using the full range of strengths of tonguing available with single-tonguing on 'dh' and 'd', and having recourse to double-tonguing 'd-g' only for passages of great rapidity. Even in the baroque period some players, it seems, advocated this approach. But the overwhelming historical evidence supports the use of a wide range of double-tonguings in flute and recorder playing.

Tonguing as described by 16th- and 18th-century writers[1]

Many recorder-players on first reading Ganassi's *Fontegara* (1535) are staggered to discover that so complex and refined a technique was used in recorder playing (at least in Italy) at this date. It is reasonable to suppose that highly developed

---

[1] See Jean-Claude Veilhan, *The Baroque Recorder*, pp. 2–20, cited in Appendix 2.

techniques had existed for many decades previously. This is especially true of tonguing. Most of *Fontegara* is devoted to the art of improvised ornamentation and in Hildmarie Peter's translation there are only five pages of technical instruction, a fingering chart, and a chart of trills. Ganassi deals with tonguing in thirty lines, but they are rich in information.

The most important point is that all Ganassi's articulations are in double-tonguing. He divides these into three basic sections, being hard and sharp, intermediate (= normal), and gentle and soft. The hard and sharp group has 't' or 'd' as its first component, 'k' as its second. Remembering that all Italian consonants are aspirated much less strongly than English consonants, this will equate with our modern 'd-g', i.e. a pairing of quick-speaking tonguings. Ganassi places vowels after his consonants to indicate the gradations of tonguing quality available on each consonant, and, I suspect, to stress his principal point that the recorder should always imitate the human voice. Ganassi's intermediate group has the slow-speaking 'r' as its reverse stroke, and 't', 'd', or 'k' (all quick-speaking) as its direct stroke. The soft tonguings, which almost 'melt into one' (i.e. become nearly a slur), are all based on 'l-r' i.e. two slow-speaking tonguings, but as Ganassi says, all this shows only 'a few of the possibilities'.

That Ganassi's advice represents standard renaissance practice is confirmed by the injunctions of other writers (e.g. Agricola, 1529) to practise 'diri-diri's. It would seem, therefore, that renaissance players cultivated light, slow-speaking legato tonguings; to play renaissance divisions portamento (on a renaissance recorder) sounds infinitely more right than using a staccato level of attack, especially as the latter detracts from beauty of phrasing. It also suggests that 'chiff', needing sharp tonguings, was more of a special effect, probably most used in dance music. The need for double-tongued legato ('dhaa-rer-lhaa-rer' or 'dhoo-ri-lhoo-ri') is all the more important in renaissance music as players were expected to play fast *passaggi* of semiquavers or quicker notes without slurring. Slurring was probably a last resort for extremely rapid tremolos; perhaps it was even judged as an admission of poor technique. Slurring was used in the seventeenth century, but at first only sparingly over pairs of descending notes.

Seventeenth-century Italian references to wind-instrument

tonguing do not vary significantly from Ganassi's precepts[1].
The more detailed accounts of Freillon-Poncein (1700),
Hotteterre (1707), and Quantz (1752) on flute tonguing give a
good idea of the art in the eighteenth century. All three writers
give musical examples with a tonguing scheme set out in full
under the notes. Freillon-Poncein makes it clear that soft
tonguings (Ganassi's intermediate) remain in the ascendancy,
even though the recorder had changed in construction and
voicing. His main concern, as ours must always be, is in the
application of the more precise 't' and the more unctuous 'r' to
various phrasing contexts, with particular reference to inequality in the French style. Although Corrette (1750) advocated
single-tonguing technique, Quantz's advice (1752) is based on
double-tonguing with 'r', but with 't' or 'd' as the direct stroke in
different contexts. He is very careful to characterize the different
uses of 't' and 'd'. Passages of single- and double-tonguing are
much more freely mixed, with single-tonguing on successions of
slower notes. Quantz says (in Reilly's translation, cited in
Appendix 2): "'Tiri" is indispensable for dotted notes; it
expresses them in a much sharper and livelier fashion than is
possible with any other kind of tonguing.' Quantz's examples,
illustrating a variety of contexts, deserve close study (they are
set out in Veilhan's *The Baroque Recorder*), remembering to scale
down the 't's and 'd's to French consonant levels, and to
recorder, rather than flute, articulation. It is Quantz who first
mentions 'did'll did'll' for double-tonguing rapid semiquavers,
a technique he derived, he says, from earlier players (it was not
his invention).

Quantz advocates the use of a variety of tonguings to avoid
uniformity of attack and to 'give each note its proper
expression'. For, he says, 'the tongue is the means by which we
... animate the expression of the passions in pieces of every
sort, whatever they may be: sublime or melancholy, gay or
pleasing'. It is tonguing, therefore, which differentiates one
player from another; '. . . these differences rest upon the correct
or incorrect use of the tongue'.

---

[1]See M. Castellani and E. Durante *Del Portar della Lingua negli Instrumenti di Fiato*
(Florence, 1979).

Tonguing in relation to speech

Even were it not for the advocacy of tonguing variety by contemporary theorists from the periods of most concern to recorder-players, the nature of the recorder itself would lead players to adopt the same approach to tonguing. Unlike any other wind instrument, the recorder requires a level of breath-pressure which is little more than that used in normal speech. It places no strain upon the lungs, nor upon the lips. One can almost 'talk into' a recorder. As Hamlet puts it, ' 'tis as easy as lying'. Because of the moderate flow of breath used in recorder playing, the instrument is extremely responsive to the articulations used in say, making a speech, or reading poetry aloud. Ganassi says that with some players it is possible 'to perceive, as it were, words to their music'. Recorder-players are failing to take advantage of the special sensitivity of their instument to subtleties of articulation if they do not use a wide range of tonguings drawn from the articulations of speech, especially those of poetry. This is particularly important in renaissance music where the subtle relationships between speech rhythms and the on-going pulse (even though the latter is weakened by polyphony) are paramount.

The relationship of recorder playing to poetry is even more instructive if consideration is given to poetry in different languages. For example, French, unlike English, is spoken without tonic stress, each syllable in a long word having much the same weight. French poetry cannot therefore be based on syllabic stress. Its gentle rhythm derives from contrasting length of syllables. In articulating French music, the duration of a tonguing is therefore more significant than tonguing stress. The slower-speaking tonguing 'r' may thus become more dominating than the quick-speaking 't', both enunciated at the same strength (it must always be remembered, however, that *all* tonguings must be done quickly and neatly – 'l' and 'r' are only *relatively* slower-speaking than 'd' and 't'). Thus in French music, 'r' may come on a down-beat, provided it is preceded by an off-beat 't' or 'd'. 't' is seen as weak in this context because it is quick-speaking. Inequality in French music (i.e. the lengthening of the first quaver in a pair and the shortening of the second) emphasizes the role of the slower-speaking 'r' which takes the on-beat longer quaver, creating the basic iambic rhythm of eighteenth-century French music, 'd – ŕ, d – ŕ'. Trochaic rhythms have their place in French music for contrast, where a more stressed 'd' takes the down-beat and the 'r' the up-beat as in normal English or Italian enunciation, but this has the effect of reducing the inequality in the pair of quavers. French was the language of *politesse* across Northern Europe and the recorder-player needs to comprehend this mode of tonguing in interpreting much seventeenth- and eighteenth-century music.

The slow-speaking 'r', articulated with plenty of stress, may be used with 'd' to set up a forward-swinging chaconne rhythm. The strong 'r' becomes almost 'th', but this is a lisped 'th' formed with the tongue upon the teeth-ridge, not on the teeth. A firmer rhythm, e.g. in a sarabande with its strongly stressed second beat, may be achieved by (exceptionally) bringing the tongue forward to enunciate 'th', as in 'the', against the upper front teeth. This rhythm may be represented 'De ther, tiDe ther', etc.

Acquiring the art of double-tonguing

It is best to begin with 't-k', the most precise and mechanical of double-tonguings, but the least expressive. It should be practised with quite a strong flow of breath on the note G'. Initially the vowel sound used should be 'i', where the tongue is high in the mouth and the traverse to and between the tonguing positions is short. To start, therefore, play loudly, staccato, and fairly quickly, a repeating phrase ♫ ♪, saying to yourself 'ticky-tee'. It is especially important not to proceed further until the tonguing is absolutely regular and balanced. Too much haste at this point will cause problems of synchronizing tongue and fingers at a later stage.

Next slacken the tonguing on the same exercise to 'dugger daa'. This is a little slower-speaking, and the tongue movement greater and more deliberate.

Now practise the same exercise with four semiquavers and a quaver, then six, then eight semiquavers, still on a loud G'. You should now be finding, however, that you can double-tongue at lower breath-pressures, and less staccato. Move on to repeated upper C's, then, as you gain control at lower breath-pressure, to lower C and other notes between.

You have now reached the point when you should try double-tonguing on two alternating adjacent notes, E to D. Be sure to synchronize the finger movement exactly with the tonguing. Try other pairs of two adjacent notes in the centre of the recorder's range which involve a single finger movement, up or down. Do not yet go below C or above C'.

Now double-tongue ♪♪♪♪ ♪ from G' to C, all the time maintaining a clockwork regularity. The patterns to attempt next are C to G', D to A', and A' to D. Having mastered these, you may now venture into other semiquaver and passage-work groupings, including whole scales, gaining confidence, speed, and smoothness. Apply your new ability to pieces you have played before, perhaps badly, with single-tonguing.

You will notice how double-tonguing helps in fast passages, e.g. in Telemann's trio sonatas in G minor and D minor. Double-tonguing has the excellent secondary effect of evening-up wayward rhythms, due to the pairing of the notes and the placing of an accent on the 'dh' note. It is invaluable in playing four isolated semiquavers nicely in time as well as pairs of semiquavers which might otherwise get rushed. Double-tonguing can smooth out a three-semiquaver entry to a phrase, which should be pronounced 'gher-dher-ghi(-dhoo)' with an accent on the 'dher' (*PB* Ex. 15): the pattern is often met with in overtures in the French style, where the groups of three demisemiquavers are articulated 'g-d-g-'. In most contexts in eighteenth-century music the time-value of a dotted note is longer than it is in modern music; conversely, the short note following a dotted note is played shorter than it is written (a useful concept is that the short note takes up one ninth of the beat in which it occurs). With the single-tonguing 'd-dhoo' this very short note tends to be muffled, but the double-tonguing 'g-dhoo' gives a clearer and tauter enunciation, and encourages a forward impetus.

You will now have reached the critical point in acquiring the art of double-tonguing, in which compound articulation must be thought of not only as a way of playing fast notes neatly, but also as a means of expression.

The starting point is to slow down the 'd-g' tonguing, playing legato and fairly softly. Keep it absolutely even and practise

♪♪♪♪ ♪ first on repeated Cs, then on lower repeated notes,

then on moderato passages involving groups of low notes: use the syllablization 'dhoo-gher-dhaa-gher'. As you get used to this tonguing at the slower speed, you will begin to find it somewhat monotonous. So substitute the slower-speaking, softer 'r' for 'gh', to provide contrast: 'dhoo-rer-dhaa-re'. Play your groupings more slowly, more legato, lightening the tonguings all the time. Finally substitute 'laa' or 'loo' for the second 'dhaa': 'laa' is close to 'dhaa' but is a little slower-speaking and softer. Your tongue will assume a kind of rolling gait within your mouth. Make sure that the gentle stress pattern of this tonguing accords with the weighting of four crotchets in a four-four bar (see p. 38). In fact you will find that the exercise of conveying this rhythm to a listener by means of tonguing only is decidedly easier with double-tonguing. This tonguing takes practice (which may be without the recorder) to achieve a regular, clean, and even articulation. Do not allow the 'r's to become too sluggish.

Play and practise 'd-r-l-r' tonguing, on a repeated low-register note at first and quite slowly, with across-the-bar phrasing – '**d** r-l-r-**d** r-l-r-**d**'. This is the standard pattern of continuous quavers in renaissance music.

Up to now, all your 'd-r-l-r' practice should have been on low notes which respond more readily to the slow-speaking tonguings. Venture into the middle and upper ranges of the recorder, but never use this tonguing on high notes because it is too imprecise (see Chapter VI). Play it gradually faster and a little louder, but never allow the tonguing to relapse back into single-tonguing (as it may threaten to). You will need to use 'i' as your vowel, entailing a shorter tongue traverse, but never lose the rolling feel of the tonguing. At last we have arrived at the standard renaissance 'diri diri' or 'diri liri' tonguing. Try applying it with a tenor recorder to the Van Eyck Variations. Again you will need much practice (abundantly supplied by Van Eyck) to master a soft legato 'd-r-l-r' at speed. Remember that in renaissance and baroque music you should only have recourse to the easier and more mechanical 'd-g' tonguings when absolutely necessary. Very fast renaissance ornamentation is best taken 'dh-y-y-y-**y**-y-y-y' (on 'i') with extremely rapid and short tongue movements – avoid slurring if possible (see p. 113). In the lower register the renaissance 'diri liri' must have smoothed out to 'doo-ri-loo-yi'.

The next phase of understanding double-tonguings should be devoted to the use of 'r' in French music. It should be practised first on strings of dotted or unequal quavers with the short off-beat note as a light 'd' (not the more aspirated 'dh' as that is too slow-speaking) and the longer down-beat note as 'r' taken neatly and lightly. Remember that 'r' must always be preceded by 'd' so it cannot be used for the first note of a phrase. 'r' was not generally used for the last note of a section, nor for a note bearing ornamentation. The best exercises are Freillon-Poncein's *Préludes* (ed. Lasocki, Faber). Avoid stress, even at the beginning of the bar: the melodic or harmonic structure of the music will give sufficient emphasis of itself.

To be able to double-tongue 'd-dl' is a valuable but not absolutely essential requirement. Follow the same procedures as for 'd-g'. Ideally, let Quantz be your teacher (Reilly translation pp. 79–85). 'd-dl' is especially valuable on pairs of semiquavers, or on an unslurred turned eight-semiquaver trill, as it is faster than tonguings with 'r', without being as abrupt as tonguings with 'g' or 'k'. With practice, it will work on high notes, though it is an extreme test of any double-tonguing to be able to use it for rapid upward leaps, for this requires a sufficiently precise reverse stroke to achieve clarity in crossing from the low to the high register while keeping rhythmic emphasis upon the low notes. 'd-dl' is extremely valuable in phrasing semiquaver scale runs, especially rising scales, when the phrasing usually is off the beat. Try a rising scale 𝅘𝅥𝅮𝅘𝅥𝅮𝅘𝅥𝅮 𝅘𝅥𝅮𝅘𝅥𝅮𝅘𝅥𝅮𝅘𝅥𝅮 𝅘𝅥𝅮 with the mnemonic 'to tootle do, de durdle dee' (i.e. Quantz's tonguings plus vowel forms to take account of the upward movement) and you may become convinced of the value of this form of tonguing. At high speed a slight side-to-side movement of the tongue across the teeth-ridge gives an additional sense of control in 'd-dl-l-dl' tonguing.

Much of the skill of recorder playing lies in accurate synchronization of tongue strokes, especially in double-tonguing, with finger movements – this is the key to virtuosity. It is tempting to give more room for the finger movements by using quick-speaking double-tonguings and short notes (staccato). This may give an impression of dazzling neatness but gives less opportunity for good phrasing. The strings of semiquavers in renaissance variations or baroque passage-work need shapeliness of phrasing just as much as an Adagio song-tune: indeed they call for more thought about phrasing because phrasing patterns are less obvious. Phrasing is best achieved with a mixture of legato and staccato to give light and shade to passage-work. The acquisition of a good 'diri-liri' as the essential basis of legato double-tonguing is therefore of the highest importance.

Unfortunately, in a very long passage of double-tonguing (e.g. the second movement of Bach's C major Flute Sonata) the tongue will become tired long before the fingers, thereby causing the tongue to lose co-ordination with the fingers. Much practice of double-tonguing, with or without the recorder, will help to remedy this, but fatigue can be alleviated if you switch from one type of double-tonguing to another. This also affords variety of attack to the music. The Bach movement I have mentioned can be 'coloured' by changes of tonguing from one section to another, tonguing serving alongside changes in volume and good phrasing of the semiquavers to draw out the full quality of the music.

Sometimes the balancing of a phrase and the avoidance of being tongue-tied can be better achieved in fast passages by using a strongly articulated 'l' as the launching pad for a series of 'd-g' tonguings. This is especially true with long passages of triple-tonguing (see below). It is the *contrast* between the slower-speaking 'l' and the quick-speaking 'd-g's that provides the sense of control: this is the whole basis of the effect of double-tonguing in evening-up passages which, played with single-tonguing, would trip you up.

Most tonguing consonants can be used as the 'reverse' element in double-tonguing if they are taken lightly 'on the rebound' with no fresh impulse of breath, e.g. 'Teddy', 'Ditty'. The relationship between tonguing and breath impulses from the lungs is fundamental to musical expression; it is this technique which underlies the shaping of a phrase, and the balance of groups of notes within a phrase.

Triple- and other tonguings

Triple-tonguing

Triple-tonguing is a necessity for triplets of high velocity, as in the first movement of Vivaldi's 'Tempesta di Mare' concerto, Op. 10 No. 1 (*PB* Ex. 16). At this speed only the fast-speaking 'k' and 'g' are of use. According to the phrase shape, use either 'd-g-d-, d-g-d' or 'd-g-d, g-d-g'. Mnemonics, however silly, help control (e.g. 'do good to gaudy girl'), and enable you to maintain a particular tonguing pattern in long fast passages without becoming tongue-tied. These triple-tonguings, aspirated with 'dg' and 'gh', or the more expressive 'diddly', are also of value in slower moving triplets or jigs, to even up and enliven phrasing.

Flutter-tonguing

Flutter-tonguing consists of trilling the tongue against the roof of the mouth in the manner of an extended rolled 'r': a fairly high breath-pressure is usually needed to get it going. It is used in Benjamin Britten's *Noyes Fludde* to imitate the cooing of the dove.

'Y' tonguing

Although there is no historical evidence for what may be called 'y' tonguing, it is a versatile device for those who wish to take advantage of it. Strong and light 'y' tonguings serve different purposes.

The strong version of 'y' may be represented as in an emphatic 'Oh, yes'. The forward thrust of the tongue is halted at its sides by the upper molars, so that the tongue is just prevented from grazing the teeth-ridge, although it comes very close indeed to doing so. This impedes the flow of air sufficiently to give an impression of separated notes. Try a forceful regularly repeated 'd-y-y-y' on a high note such as C', slowly at first to get the muscular feel of this tonguing. When you have a regularly repeated 'y' tonguing under control on a high note, work down to lower notes, with as much deliberation but less force. Some instruments (especially basses, which abhor strong tonguings), may articulate better in the low register with 'y' rather than even a soft 'dh'. 'y' is unsuitable for articulating a note in the high register because it it too slow-speaking. Scale practice with 'y' is salutary in that it evens out irregularities that may be perpetrated in slurred scales. 'd-y-y-y' is useful for enunciating and balancing scalic four-semiquaver sequence patterns in baroque music. A good practice piece is the last movement of Vivaldi's 'Cardellino' concerto, Op. 10 No. 3. This includes repeated groups of unslurred semiquavers on B'-$C\sharp'$-D'-E': many recorders will slur from D' to E' without crossing the register break by lifting 2, and with 'y' tonguing this fingering may still work yet give the semblance of unslurred semiquavers.

The light version of 'y' may be represented by the unemphatic 'y' in the phrase 'Do you take tea?' The tongue is now only lightly touching the molars and is a tiny bit further back from the teeth-ridge, though close enough to constrict the flow of air sufficiently to differentiate the phrase from 'Do take tea'. This is the lightest possible of all tonguings, and it can convey the impression of a slur if done artfully enough. Though light, it has to be precise, and in perfect synchronization with finger and thumb movements: it should never become flabby. The regular, slight (but definite) movement of the tongue to and from the light 'y' position will even up difficult slurs, such as fast chromatic runs. It helps, as a 'touch of tonguing' ('doo-yee'), in slurs across register breaks, as the constriction made by 'y' causes the increased air velocity needed for the upper register. Try its effect in the example on p. 40, and in Exx. 32–44 of the *Practice Book*.

A strong 'y' may be found useful in separating the eight notes of a very fast turned trill in renaissance ornamentation at the end of a 'groppo', and a light 'y' in controlling slurred baroque trills. Try (at speed) the trills on p. 116, first separately tongued with strong 'y's – 'd-y-y-y-**y**-y-y-y-d', then with light 'y's controlling the balance of a slurred (baroque) trill, and then slurred without 'y's.

Articulations without tonguing

It is essential for recorder-players to be able to play a note by, as it were, direct blow, without tonguing. This enables a note, especially at the opening of a piece or phrase, to start unobtrusively and then develop and fill out. As Ganassi says, 'imitate a singer'. An excellent way to acquire this form of enunciation is to play Schubert's 'Who is Sylvia, what is she' (starting on A') matching the delivery of a singer. 'Who' and 'what' need a soft and deep 'h' enunciation, the air-stream welling up spontaneously from the lungs and through the throat.

Ending a note

Ordinarily, there are no difficulties. The return of the tongue into the correct 'dh' position effortlessly and efficiently both stops one note and prepares for the next, during which tiny stoppage of airflow all finger movements take place.

Difficulties arise when there is no next note. They mount still more if the last note of a phrase, or worse still of a whole piece, happens to be a long note. If this last long note is marked *ff* the odds are that the player will swoop up an exuberant semitone in a final burst of defiance to the composer, and if it is marked *pp* he will expire a semitone flat with appalling pathos. Control must be retained until the very end of a piece of music – and a second or two more. If the last note makes demands on the lungs, a little vibrato may help to keep it steady. The actual stopping of the note may be done in the normal way of putting the tongue back to the 'dh' position with great speed and delicacy; the effect is of the final 'n' in the elongated whispered last syllable of 'London'. If the note is very quiet and therefore hard to control, the tongue may come forward into the 'lh' position while the note is still being played, the stop then being made by a quick forward pressure of the tongue. A more outlandish way is to cut off the air with the lips, plucking the recorder from the mouth at the same time. The lips should close firmly and instantaneously behind the instrument. This method is surprisingly effective, both aurally and visually.

The imperceptibility of good tonguing

It is a great irony in recorder playing that, except for special effects such as echo tonguing or flutter-tonguing, the more you acquire skill in using the whole range of single- and double-tonguings described in this chapter, the less apparent it will be, as a technique, to the uninitiated listener. You should practise double-tonguings until they are as neat, regular, and as even as single-tonguings. What you and your audience will notice is that your performance is more varied, more lively, above all more expressive. Ultimately, tonguing technique should be imperceptible, as it becomes totally subservient to good phrasing and imaginative interpretation. If a recorder-player's performance does not hold your attention, it is probably his tonguing technique that is at fault.

IV

INTONATION

A note is in tune when it bears a perfect relation to the notes preceding and following it (melody) and to other notes being played at the same time (harmony). A person can only achieve an understanding of the relationship of one note to another by listening to music in such a way that he hears all the separate strands of a piece of music (preferably by concentrating on the middle voice and remaining aware of the treble and bass) and is at the same time conscious of the construction of each chord. A recorder-player can only keep in tune, therefore, if as well as playing his own part he listens to everyone else's. If his part is too hard or the structure of the music too complex for him to do this, he should at least listen to one other part, and that should be either the bass or the part next below. Intonation is always improved in a consort if the players sit in the order of their parts. Only by listening carefully along and through a piece of music as a whole can such justness of intonation be achieved as the narrowing of the semitone between a leading note and its tonic, or the flattening of minor and sharpening of major thirds and sixths. It is important that a note or a chord should be imagined in the mind's ear before it is played; no true musician will embark upon playing a note without knowing what it is. This chapter, therefore, describes the means at a recorder-player's disposal of communicating the exact note he has in mind.

Whatever beginners might learn, the thumb and first three fingers of the left hand on the treble recorder do not necessarily produce a definite note C. If the instrument is 'in tune' and warm, and if the player uses medium breath-pressure, he will, however, produce one of the many notes that are conveniently represented by a mark on the second space down on the treble stave, rather than one that would be better represented by the same mark with a sharp or a flat in front of it. The player has the power to cover the whole range of notes that are expressed by this C, starting from the territory uneasily shared with B right up to the foothills of C♯. He has the freedom of the singer or the violinist, and is not confined, like the pianist, to pressing a key and taking what comes.

The experiments in breath-pressure advocated in Chapter II will have shown that on the note G' the range of intonation is, on most recorders, a whole tone or slightly more. The intonation range of notes above or below G' is rather less, but even on top F‴ or bottom F most recorders will range over a semitone. The fact that recorder intonation is so sensitive to changes in breath-pressure is both an advantage and a disadvantage. It is an advantage in so far as small alterations in breath-pressure will move the pitch of a note slightly without a significant effect on the volume of sound produced. This enables the player to make minor corrections to the pitch of a note either to keep in tune with other players or to overcome imperfections natural to his own instrument. A treble player might, for example, find that his F' was a little flat and would correct it by blowing harder. One comes to make these modifications automatically when the idiosyncrasies of a particular recorder are known – indeed they must be made on every recorder for it is neither possible nor desirable that a recorder should be constructed absolutely in tune throughout its chromatic register with a stable breath-pressure.

This admirable flexibility of intonation reaches its limits when changes in breath-pressure designed to keep notes in tune have to be great enough to affect volume: it is obviously musically undesirable that a player with a very flat F' should blurt it out loud (but in tune) every time he comes to it. Conversely, the sensitivity of intonation to breath-pressure causes problems when stronger variations in volume are called for by the composer or by the style of the music than those with which your instrument can cope without going out of tune. True echo effects, for example, are delightful, but not if the 'echo' is a semitone flat. Baroque-voiced instruments have the flexibility to cope with a wider range of dynamics before going out of tune, but if the instrument is badly tuned by its maker in the first place this stability of intonation in itself causes problems.

Shading and shade-fingering

The chief technique for flattening a note is called 'shading'. To shade a note means to lower the unused fingers over the open holes until the note is flattened to the extent desired. As an experiment play the note E on the treble loudly and lower the second finger of the left hand over its hole until the stream of air coming from the hole can be felt on the ball of the finger. Now, lowering the finger slowly, press the column of air down into the hole until the finger is just grazing the edges of the hole. Very gradually press the finger home to complete your slide down from E to D. Shading with the uppermost unused finger is extremely critical as a tiny movement affects the note's pitch, and it is less nerve-racking to impinge upon the lesser columns of air emerging from holes lower down: the whole process of shading E with the third finger of the left hand cannot lower it as much as a semitone, and the first finger of the right hand has less than a quartertone effect. It is perhaps best to shade with all the fingers at one's disposal. Try playing E and moving the unused fingers up and down to produce a controlled, slow, wavering effect reminiscent of American railway engine whistles. In this exercise the fingers of the right hand may actually cover their holes, but the two left-hand shading fingers have to be moved with more care and should never be low enough to touch the instrument. Another method of shading is to place the shading finger on the instrument but at the side of its hole, from where it may roll over towards the hole as required. This method is particularly useful when only one finger is available to do the shading; an example is bottom G which is sometimes a little sharp and can be flattened by placing the little finger, politely bent, on the brink of its half-hole. Middle G′ is another note that is often sharp, and although it can be shaded from above, the easiest way to flatten this note, as well as F′ and alternative E, is to swivel the left wrist back so that the first finger of the left hand leans against the side of the instrument, with the finger almost straight and jutting out diagonally above the hole in such a position that bending its middle top joint causes extra flattening. The other fingers lie low, and the little finger may actually be on its hole covering both half-holes. Another method of shading, which seems to suit best the fingers of the right hand, is to arch the finger across its open hole to touch, or nearly touch, the edge of the far side of the hole.

Shading may be achieved by covering holes left open, provided that the hole taking the shade-fingering is low down on the instrument in relation to the position of the fingers making the note itself. The most useful fingers for this method of shading are the little finger and the third finger of the right hand, for the degrees of flattening they control are made finer by their double holes. Little-finger shading is the most delicate of all: shading a note such as E or F' with the little finger scarcely lowers the pitch, but the little finger comes into its own not only on lower notes such as A or C but in the upper octave where the effects of shading are accentuated (compare the flattening effect of adding the little finger to the ordinary A' fingering with the same movement an octave down). Another form of shading is the covering of a hole below an ordinary 'forked' fingering. Many recorders, for example, produce a sharp C♯ with the usual fingering (0 12– 45– –), and the third finger of the right hand has to cover its half-hole or both half-holes to bring the note into tune. Alternatively, if the sharpness is slight, little-finger shading may be applied. Another method is to lower the third finger of the left hand into the air-stream emerging from the open hole beneath it to produce the same flattening effect; control may be gained by resting the little finger of the left hand against the side of the recorder. The choice of method depends on which comes easiest to the player, though it is generally advisable for full shade-fingerings to be applied to notes that are consistently out of tune, while the other methods of shading are used for controlling intonation in passages where dynamic range and expressiveness are called for, and for corrections resulting from intonation exigencies of ensemble or consort playing, or the use of remote keys. Shading is vital when whole passages need to be played loudly, in order to avoid sharpening.

It comes as a shock to realise that this technique is as old as Ganassi – 'An instrument can imitate the human voice by varying the pressure of the breath and shading the tone by means of suitable fingering.' (*Fontegara* Chapter I).

Slide-fingering

Sharpening a note without increasing breath-pressure may be achieved by 'slide-fingering'. This simply means pulling the lowest of the fingers forming a note to one side to expose some or all of the hole it covered. It is easy to do this on a note that is normally a forked fingering, such as B♭ or E♭: it is simple, too, with most of the 'pinched' notes of the upper register, for if the

thumb-hole aperture is widened, the note is slightly sharpened –
but care must be taken not to overdo this, otherwise the
octaving effect of the pinched hole is spoilt and the note breaks
downwards with an ugly crack. Intense concentration is
needed, however, if a plain-fingered note such as C is
sharpened, for a hair's breadth moving of the third finger from
its hole will send the note up. Control is enhanced if the
operating finger is pressed firmly on to the instrument so that
every edge of the hole except the fraction being released is felt as
the finger is dragged sideways. Alternatively, the finger may be
moved upwards, that is, lifted so as to rest lightly on the hole
rather than properly cover it, but this method, even more than
that of slide-fingering proper, adversely affects tone-quality –
the good round sound of the plain-fingered note is weakened as
the pitch rises. To make matters worse, in slide-fingering a
plain-fingered note such as C, tone deterioration happens quite
suddenly the moment the finger begins to cause leaking of air
from the hole. Slide-fingering plain notes is therefore a device to
which one must have recourse only when absolutely necessary.
On the other hand, slide-fingering forked notes is a vital
technique for achieving flexibility in recorder playing.

Half-holing techniques, referred to by Ganassi, were
apparently commonly used in sixteenth- and seventeenth-
century recorder playing. Ganassi writes that holes 'should be
half closed a little more or a little less according to the demands
of pure intonation . . . Remember that you can sound every note
softly by slightly uncovering a finger hole and using less breath
. . . You should half-close the holes somewhat more or less as
your ear requires and as you feel to be right.' Blankenburgh in
1654 even goes as far in his fingering chart as to prefer
half-holing to cross-fingering for normal sharps and flats,
indicating whether a hole is to be rather more than half closed,
or rather less than half closed. This may be a legacy from
transverse flute fingering (the renaissance and early baroque
flutes had no keys) as cross-fingering has a more subduing effect
on flute tone than it does on that of the recorder. On a wide-bore
recorder the tonal change on slide-fingering C is a little less
marked and sudden than it is on a baroque recorder, and
half-holing techniques, in which a really skilled player can
achieve a very fine control over intonation, can be used more
readily.

It is of course in the converse of making a note sharper by

slide-fingering with steady breath-pressure that the use of slide-fingering for dynamic control resides. If you play a note on normal fingering at *mf* and then wish to decrescendo to *pp*, it will stay in tune only if slide-fingered.

One form of slide-fingering that works well on upper octave notes without affecting thumbing is to slide-finger with the first finger of the left hand. Do this by pivoting the first finger very slightly off the far side of its hole, while still keeping the finger on the recorder ('leaking-finger' technique). This has effect down to Bb', which as a forked fingering is easy to slide-finger with 6. By using alternative A' 0 123 4–67 one can slide-finger 7 and thus complete the repertoire of *pp* notes in the upper octave. A problem with slide-fingerings is that, to a greater or less degree, in slightly sharpening the note in question they cause change to the breathy undertone that is present as a 'shadow' to the main note on most recorders and is a constituent of its tone-quality. The undertone is likely to be more in evidence on the upper notes of the second octave, and thumb or first-finger slide-fingering of high notes can cause poor tone-quality in the transition from *mf* to *pp*. You must see how it works on your own recorder.

Below A' the two-finger forking of F♯', Eb, C♯, and B all lend themselves to slide-fingering, and there are forked alternatives for the plain-fingered notes to open up slide-fingering possibilities, at least down to G (see next chapter). The half-holed low notes Ab and F♯ slide-finger without problem. Leaking-finger techniques can be tried on notes in the lower octave, normally using the finger first above the lowest finger down. This just about works with F, but G is so rich in tone that any leak affects it too suddenly.

An extreme of leaking-finger application to the bottom octave is to use the thumb. Try this at low breath-pressure from F' down the scale, checking intonation with normal fingerings at normal breath-pressure. The thumb-nail has to tighten up to its extreme as you approach the bottom of the scale, and for the last four notes breath-pressure has to fade out to almost nothing – it is extremely difficult to control evenness of breath at this very low level. This is the so called 'harmonic' register, a very quiet reedy sound of dubious quality. While it cannot have much practical value in baroque music, except for echo effects, the sound is used by avant-garde composers, and to play two octaves of 'harmonic' and slide-fingered notes *ppp* followed by the same *ff*, heavily shaded, is a good exercise in intonation control as well as a striking display of the recorder's dynamic extremes (see pp. 95–6).

Quiet notes in the lower register which are of acceptable tone-quality can be produced by using a variant of thumb-leaking. This is achieved by lightly placing the tip of the thumb-nail not within the circumference of the thumb-hole but directly on to the rim of the thumb-hole, or slightly outside it, and then allowing a 'hair-leak' to occur at the thumb-hole rim (this can be sensed by the skin of the thumb as a tiny vibration). One has to find the exact point, which lies between full coverage of the thumb-hole and the slightly wider leak which produces the hazy 'harmonic' sound, where the note plays softly, is in tune, and has a good enough tone-quality for soft playing. This should be done for each of the notes of the lower octave, the amount of leak seeming to approach infinity as one gets down to bottom F. Another way of achieving these minute thumb-leaks is to drag the thumb slightly sideways from its normal closed position, using the resilience of the surface of the skin to make the leak.

A way of tackling a full decrescendo is to choose a sharp alternative fingering as one's starting point, shade it for the *mf* at about or a little above normal breath-pressure, and to die away by gradually lifting the shade-fingerings to match the decrease in breath-pressure.

Re-fingering

Some notes may be played sharper or flatter by re-fingering them – in other words using one of the alternative fingerings that are dealt with in detail in the next chapter. To illustrate the possibilities and complexities of intonation control, let us consider the note F'. Its normal fingering (i.e. thumb and second finger of the left hand) is often a trifle flat. It is easy to make it flatter either by shading with the first or third fingers (or both) of the left hand or by using any of the twelve feasible shade-fingerings available in the right hand. It is not, however, an easy note to sharpen, for slide-fingerings in this position are almost impossibly critical. But there are plenty of alternative fingerings for F′, with and without the thumb. Here are some of them:

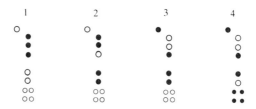

1 and 3 are the important alternatives. 1 varies considerably from one instrument to another, but generally gives either the same pitch as the normal fingering or is slightly sharper. It can, of course, be flattened by shading either with the thumb (too critical to be safe) or the first finger of the right hand, or by shade-fingering, the most useful being with the second finger of the right hand. As it is a forked fingering it is amenable to sharpening by slide-fingering. If 1 is flat, however, the variant of it shown in 2 can be used, and this, being a double fork, is extremely easy to sharpen further by slide-fingering if necessary. 3 may be a little flat, but, because it is what one may call a 'wide fork' (there being two middle holes left open), it is less critical to slide-fingering and one, or even both, of the right-hand fingers may be taken off without sending it up to F♯′. 4 bears the same relation to 3 as 2 does to 1 and is therefore a little sharper than 3. With experimentation you will find at least thirty fingerings (excluding half-holes) that produce some version of the note F′. Most of the middle notes of the recorder behave similarly. Even though low and high notes offer less opportunity for variations of fingering, intonation control of some kind is obtainable with every note on the recorder. A thoroughly bad recorder, therefore, can be played perfectly in tune, though the result would not justify mastering the difficulties.

Tuning: devices and gadgets

There is no point in knowing how to play in tune if one's instrument is basically out of tune with the other instruments in an ensemble. The first requirement of a consort is that all the recorders should be warm. The motion of sound waves becomes more rapid in warm air, and the resultant rise in pitch is not compensated for by the expansion of the instrument as it gets warmer with the player's breath. A cold recorder, therefore, plays flat.

As it is easier to flatten a note than to sharpen it, the obvious rule for a recorder consort is to tune to the flattest instrument, warmed up and at medium breath-pressure. If you know that your F′ or D is generally on the flat side with the fingering you normally use for it, and you then find it is flatter than the other recorders you are playing with, you must either re-finger your flat notes, sharpen your instrument or persuade the other players to flatten theirs.

Flattening a recorder is easily done by 'pulling out', that is, adjusting the main joint so that it is not quite home and the instrument is thereby lengthened. A quarter of an inch is the utmost practicable limit of pulling out a treble recorder as the effect is more marked on the notes of the lower octave than the upper and more pulling out causes poor intonation. If bottom F tends to be sharp, try the effect on it and on other low notes of pulling out the foot-joint by no more than a quarter of an inch (otherwise your little finger may not reach its holes). The flattening effect of Plasticine 'wings' is a little more evenly distributed, and the device can be used to tune a recorder made in one piece (e.g. a sopranino). The wings are vertical extensions by about an inch of the side walls of the 'window' of the recorder with pieces of Plasticine rolled and flattened and pressed on each side of the 'window'. The 'wheelbarrow' sound-projecting device (see p. 98) has the same flattening effect. Both devices make tone more 'plummy'. This method of tuning is used with metal organ-pipes which have two strips of metal standing out on either side of the mouth and at right angles to the lip: if bent inwards they lower the pitch and if bent outwards they raise it.

A recorder may be sharpened evenly throughout its range, with only a slight coarsening of tone-quality, by the use of tuning holes (see Kottick *Tone and Intonation on the Recorder* pp. 21–2). One or two small holes of a sixteenth of an inch diameter may be bored in the sides of the instrument at about a half to one inch lower down the instrument than the edge (more for bass, less for descant). Normally these holes are kept closed with wax (candle wax is easy to apply and remove) or with a wooden stopper, but if it is desired to raise the pitch of the instrument one or both of the holes may be opened. Amateur carpentry on recorders may, however, cause damage, so such modifications are best carried out by an instrument repairer.

If your recorder is slightly out of tune with itself, it may be quite an easy matter to re-tune it. Recorders may in fact slightly change in tuning as they age, irrespective of how well they are maintained. Again I refer readers to Kottick's *Tone and Intonation on the Recorder* (pp. 22–7) for a clear account of what can be done with the minimum of risk by a player with no pretensions to carpentry, and of how to re-tune in order to remedy various faults. However, a good ear and a second opinion are strongly recommended.

Beats and difference-tones

Recorder-players are lucky in that the 'pure-tone' quality of their instrument provides them with the assistance of audible 'beats' and 'difference-tones' to check intonation. Beats, which are most noticeable in two-part playing on high instruments, are caused when two recorders are not quite in unison: the beats become faster as the notes get more out of tune, and when the two notes are a minor third apart the frequency of the beats has become great enough to form an actual note, or difference-tone, which should be in harmony with the notes producing it. Two descant players could, in theory, play a trio creating their own bass part in difference-tones, but they would have to play most consummately in tune – and there is no reason why they should not.

V
ALTERNATIVE FINGERINGS

If the question were put 'what is most important in recorder technique – breath control, articulation, or fingering (including thumbing)?', one would prefer not to answer because all are so important, but if forced, the answer would have to be 'fingering'. Because the fixed windway and edge minimizes control by the mouth and lips, a recorder-player depends far more than other woodwind players on fingering devices to achieve good intonation, wide dynamics, and variety of tone. They are also needed to overcome the weakness of many recorders in making extraneous sounds ('clicks') when slurring across register breaks at G' to A', and D' to Eb'. As recorder semitones are cross-fingered, unlike those of fully keyed woodwind instruments, normal fingering on the recorder is rather complex, and attention needs to be given to means of facilitating passages involving contrary finger movements.

Daniel Waitzman, in *The Art of Playing the Recorder* (p. 79), challenges the idea of 'normal' fingerings – 'The concept of "standard" fingerings is of doubtful value for all woodwinds; it is of virtually no value on the recorder, an instrument with a very rich vocabulary of fingerings. On the recorder more than any other woodwind, selection of fingerings influences volume, timbre, and other qualities generally subsumed under the heading of "expressivity". Although there are no "standard" fingerings in the broadest sense of the term, there do exist fingerings which are more or less standard for a given musical-technical situation. The selection of the proper combination of fingerings to achieve the desired result constitutes the grammar of recorder fingering; it is this, rather than the mere unorganised knowledge of scores of different fingerings, that represents a true understanding of fingering on the recorder.' 'Normal' fingerings, will, however, on a well-tuned treble recorder be used perhaps nine times out of ten in playing renaissance, baroque, and much twentieth-century music (see for example the number of instances alternative fingerings are marked in the *Practice Book*). It is the ability to use the alternative fingering in the right place that distinguishes the experienced from the inexperienced player.

The physical mechanics of moving fingers and thumb on and off their holes, including subtleties of thumbing, are dealt with elsewhere in this book (e.g. p. 125) and in *Introduction to the Recorder*. Many examples in the *Practice Book* are designed to develop a good fingering (Exx. 18–49). This chapter is therefore devoted to a consideration of alternative fingerings and their use.

It is common in the process of learning, to become intoxicated with some new-found device, but excess leads to wisdom, and in more sober judgement the real value of the new piece of knowledge becomes evident. This happens with alternative fingerings on the recorder. The discovery of their existence opens up wide vistas of new possibilities of the instrument: fast passages that tangled the learner's clumsy fingers suddenly become easy; ugly slurs full of 'clicks' and 'in-between notes' become neat and smooth. New alternatives are discovered that make awkward bars into child's play. But then the player becomes more critical. Tonal inequalities in his alternative-fingered notes begin to make themselves heard, faults in intonation become apparent. He finds that the new facility in fingering which the use of alternatives has given him makes it quite easy to play previously hard passages with ordinary fingerings, which anyhow are more reliable both for sight-reading and under the strain of performance. And eventually he hits upon the paradox that alternative fingerings make the recorder harder to play, not easier. He uses them more sparingly, but to far greater effect.

The diagrams on p. 62 for F′ fingerings illustrate the principle of finding alternatives. Most notes of the F major scale are 'plain-fingered', in other words, working from the thumb down you have a number of holes all covered followed by the remainder of the holes all left open. F, G, A, C, D, and E are plain-fingered, while B♭ and F′ are forked fingerings. Nearly every plain note has a forked alternative, but not necessarily perfectly in tune with it. Thus E can be played with the thumb and the first finger, or with the thumb and the second and third fingers; this latter is the best known of all alternative fingerings. Some of these forked alternatives can be 'double-forked' by lifting the lowest finger of the forked fingering and putting down fingers below it. Many forked fingerings can be widened by leaving two holes open and counteracting the sharpening by covering more holes lower down. An ordinary forked fingering can be so much flattened by adding fingers below that a note a semitone, a tone, and possibly even a tone and a half below the original can be made to sound, so giving alternatives for other notes. Always remembering that a wide open thumb-hole constitutes a forked fingering, this principle works right up to D′, above which forking sharpens a note instead of flattening it – consider, for example, the normal fingering for E′. Alternative fingerings produce notes that vary to a greater or lesser degree in intonation and tone-quality from notes produced by normal fingerings. Generally speaking, the more forked a note the poorer is its tone-quality and its capacity to react to stronger tonguings: even carefully controlled vibrato cannot disguise its tonal weakness. Forked fingerings over an open thumb-hole tend to coarseness.

The tonal differences between a plain note and an alternative are seldom so slight that they can be ignored. In the following uses of alternatives, therefore, these variations in tone-quality have either to be overlooked or exploited.

1. Trills

In trills and other rapid decorations the notes pass by so quickly that their tone-quality is hardly apparent. Neatness and shapeliness are the main factors. Good execution at speed can best be achieved by utilizing the minimum finger movements in passing from one note to another, and the employment of an alternative fingering can sometimes make the difference between moving five fingers or more and moving only one to change a note. This, then, is the most important use of alternative fingerings – indeed, they are often called 'trill-fingerings'. Sometimes the tonal inequality of the two notes of a trill adds interest and brilliance but care must be taken not to allow a preponderating note to spoil the pattern and fluency of a trill. Because a trill-fingering is generally forked, it is usually easy to add the turn at the end of the trill by putting down fingers below the fork: with normal fingerings trills with turns are often unmanageable. The forked fingering of a trill often makes the final turn, which would otherwise threaten to be a mess, very simple to play (usually two fingers below). It is frequently unwise to begin, and, even worse, to end a trill in an alternative fingering position, for once the repercussions have ceased, the possible poor quality of the note shows through. Quick thought matched by quick finger movements is necessary to get from the plain-fingered appoggiatura on to the alternative for the trill itself, and then off it for the closing note.

2. Slurs

Quick slurs often demand alternative fingerings for the same reason that trills do. Try the effect, for example, of slurring G′ A′ Bb′ quickly with ordinary fingering and then with the usual alternatives for G′ (all on except thumb) and A′ (lift second right): one is inevitably clumsy and full of 'clicks', the other is – or should be – neat. But this slur in a slow, quiet movement cannot be alternatively fingered as the coarse quality of the G′ followed by the much thinner A′ spoils the evenness of the flow, and ordinary fingerings with their better tone-quality have to be used (see p. 52). Somewhere between adagio and allegro, varying according to the player's skill in tonguing and his instrument's voicing and tone-quality, lies the point where the danger of perpetrating a 'click' matters more than the disadvantages of unequal tone, for in all slurs, and in particular wide slurs, the player must above all aim at a smooth transition – that is the composer's intention in marking a slur. Making 'clicks' over slurs is the recorder's equivalent of bad pedalling on the piano.

3. Runs

Slurred runs in quick music can seldom be played neatly without alternative fingerings: chromatic runs are nearly impossible without them. Moreover, a run which threatens to be uneven because of a difficult finger change immediately before an unaccented note can be made fluent and shapely by using an alternative fingering which throws the greatest finger movement on to an accented note. A good example is the D major scale played in the usual pattern of quaver, six semiquavers, crotchet (*PB* Ex. 42). The accented note in this run is G′: with normal fingering the greatest finger movement is from G′ to A′, which throws an accent on the A′ and makes a 'click', but if the G′ alternative is used the unavoidable five-finger movement is from F♯′ to G′, where the accent is needed. Incidentally, the C♯′ at the end of this slur can be played with the second finger of the right hand instead of the first so that a one-finger movement (lifting of third left) is substituted for the normal three-finger movement. Economy in finger movement is the main purpose in using alternative fingerings in fast music.

4. Passage-work

The playing of broken chords and other semiquaver passages in allegros is simplified by alternative fingerings. Particularly useful are fingerings such as F′ with three left and no thumb (illustrated on p. 62), and G′ with third left and first right only, as they can form a pivot around which other notes may be grouped. For instance, the common semiquaver groups A′ F′ C F′ and G′ E♭ B♭ E♭ are difficult to play rapidly with ordinary fingering, but with the alternatives mentioned they become easy because the fingers forming the alternative remain still for the whole group of notes, enabling movements higher and lower on the instrument to pivot round them with a sort of rocking motion: the finger(s) forming the 'fulcrum' should be held down firmly. This rocking effect is the secret of well-balanced passage-work. Intelligent fingering prepares the way to good interpretation.

5. Intonation Control

This use of alternative fingerings was dealt with in Chapter IV. A player should experiment with his instrument to find to what extent he may use alternative fingerings for intonation control without unduly affecting the evenness of the tone-quality produced.

6. Decrescendos

All alternative fingerings are forked fingerings; in the last chapter it was pointed out that forked fingerings are amenable to sharpening by slide-fingering. Conversely, such slide-fingerings may be used to keep intonation constant under a variable breath-pressure, so allowing a note to swell or die away. The latter effect is difficult to manage on plain-fingered notes and an alternative must generally be used, but once again with the proviso that it does not unduly change the tone-colour, or, if it does, that such a change is justified.

7. Tone-quality changes

The different timbre of alternative-fingered notes may be exploited to add to the range of expression of the recorder (see Chapter VIII).

8. Keeping within the register

There are fingerings both with and without the pinched thumb-hole for A′, Ab′, and G′ (and also F♯′ but it is rather stifled). The timbre of the pinched notes matches that of the upper register, while the open notes have the fuller tone of the lower register. In playing a slurred adagio phrase that goes up or down to one of these notes it is as well to use the fingering that is least likely to make the note predominate by sounding different in quality. In fast music these notes are valuable to effect a smooth transition between the registers as the thumb may be brought into position while they are being played (as, for example, in the common G′ A′ B′ slur – *PB* Ex. 39).

In the following survey of alternative fingerings allowance must be made for differences between instruments: an alternative that works on one instrument will not necessarily do so on all others.

Alternative fingerings in the lower register

E: Officially the lowest note of the treble recorder is F. But a low E can be faked by lowering the instrument against the upper leg so that the bottom opening of the instrument is sufficiently shaded to flatten F by a semitone; the note should be played with little or no tonguing or it will sound F′. This, with care, can be surprisingly convincing, so long as nobody notices.
F: There are no alternatives for F (but see p. 61). In order to

play it pianissimo, gingerly slide-finger 7 over the bottom half-hole. F easily goes flat with lowering of breath-pressure. For a loud F, shade the bell as for E, and use light, slow 'r' tonguing.

F♯: G♭: No alternatives, other than 'harmonics'. The 'bending finger' and 'wrist swivel' methods of covering half-holes are described and illustrated in *Introduction to the Recorder* pp. 43–4. F♯ can be made more powerful by bell shading.

G: A rich-sounding note, slightly sharp on some recorders. If so, prefer reduction of breath-pressure to shading, as shading is of necessity very critical and coarsens tone-quality.

In the trill A♭ to G, G may be played 0 123 456̸7*, with 6 well arched and without swivelling the right hand round to the half-hole position; trill with 7 (asterisked).

A♭: G♯: 0 123 45–7 with 6 heavily shaded may be used as the only alternative to half-holing.

A: An alternative which provides one way of trilling B♭ to A is 0 123 4̸567, with 5 trilling in a well-arched position on to the near half of its hole. Placing the little finger of the left hand on the side of the instrument will give extra control over finger movements of the right hand.

B♭: A♯ Sharper (or softer) versions are 0 123 4–6̸7, 0 123 4–6– 0 123 4–6̸7, and 0 123 4–6̸–. The latter is useful for the B♭ to A♭ trill, trilling with 5.

A very slight lifting of 5 from the normal A fingering will give a poor B♭, virtually a semitone sharpening of A by slide-fingering. The right hand should be raised so that 5 is in an arched position and the curve of the fingertip just touches the far side of the hole, leaving the near side open. Although the tone-quality of this note is weak and stifled, it provides another way of doing fast trills from B♭ to A. The trilling finger continues to touch or nearly touch the recorder during this trill, seeming to move on the elasticity of the fleshy finger-pad. Although the movement is small, it should be controlled and regular, otherwise poor intonation and blurring result. The recorder needs to be held firmly to negotiate this trill, so it is helpful to steady the instrument by putting the little finger of the left hand underneath it, pressing upwards. Both this and the B to A trill (see below) tend to sharpness as it is difficult to keep the finger movement narrow enough. Unless the trill has to be played very rapidly it is better to do a B♭ to A trill with normal fingerings, despite the difficult contrary finger movements in the right hand: 7 is not necessary for this trill.

B: If normal B is sharp, *7* should be added.

Normal B, being a two-finger cross-fingering low on the instrument, tends to be weak on larger recorders, and has to be given as full a tone as possible with the right tonguing and breath-pressure, and perhaps (depending on the context) a touch of vibrato.

The sharp alternative 0 123 –5–– may be used on the common B to A trill. Sharpening makes a trill more brilliant, but it is a trick that can become a vice (though Ganassi advocated it). B to A trills can be more accurately negotiated with 4 moving in the same way as 5 does for the B♭ to A trill. It is preferable, however, to play this trill with normal fingerings, despite the contrary motion.

B to A♯ trills are done with 4 trilling on half its hole.

The poor-quality alternative 0 12ƶ 4567 may be used for the turn after a D to C♯ trill.

C: 0 12– 4567 gives a wheezy alternative that has some value in soft passages and does service in the rare C♯ to B♯ trill – it occurs in a Bach flute sonata – and as a D♭ to C trill in a Boismortier trio (*PB* Ex. 58).

0 1–3 4567 may be used for the turn after an E♭ to D trill, but it is almost as awkward as ordinary fingering.

C♯: D♭: It is debatable whether the normal fingering for this note is 0 12– 45––, 0 12– 456–, or even 0 12– 45ƶ–. Some recorders have a C♯ key.

A C♯ to B trill is taken 0 123* –56–, with 3 trilling low (*PB* Ex. 26).

For the C♯ turn after an E to D♯ trill 0 1–3 456*7* may be used, and for a D♭ turn after an F' to E♭ trill 0 –23 45ƶ–: both these alternatives are too poor in quality for other uses.

D: The most important alternative fingering to be dealt with so far is 0 1–3 45ƶ–, the 'one-and-a-half-below' alternative for D. This is used primarily in the E♭ to D trill and slur. It has a thin and remote quality that contrasts strongly with the powerful ordinary D, and as it is heavily forked it is amenable to slide-fingering and therefore to a decrescendo effect. This alternative is a little on the sharp side, but as D is flat on some instruments by ordinary fingering, the one-and-a-half-below fingering is invaluable for getting it in tune when a compensatory increase in breath-pressure would be undesirable.

The value of this fingering in the downward slur from E♭ may be demonstrated by playing the slur with the ordinary and then the alternative fingering: the transition from the subdued tone-quality of the forked E♭ to the firm ordinary D is ugly compared with the gentle sighing effect of the slur to the alternative D. For E♭ to D trills 6 may remain down so that only 5 is trilling.

On some recorders the one-and-a-half-below D is too sharp to use, and the alternative must be fingered 0 1–3 456–. This makes the C turn after an E♭ to D trill easier, using 7.

Another useful D alternative is 0 –23 45––. It is too heavily forked to be of good quality; its tone resembles that of the one-and-a-half-below alternative but is wheezier. It comes into its own on trills and fast slurs over the notes F′–E–D, where the well-known 0 –23 –––– alternative is used for E. Whenever the notes F′–E–D–E are encountered in fast music, this fingering should be employed.

The D alternative – 123 456– is of poor quality, so is only serviceable as a turn after an F♯′ to E trill – 123* 4–6–. 0 ––3 4567 is only a tonguing exercise (see p. 34).

E♭: D♯: As the normal fingering of E♭ is forked it offers all sorts of variants: adding fingers below (7 variants) flattens it, while double-forking, e.g. 0 1–3 ––67, generally sharpens it. Because the ordinary fingering has a tendency to sharpness the most useful of this large group of alternatives is the flattening 0 1–3 4–6–.

A second group of nine possible alternatives is based on the important fingering 0 –23 4–––. This note has a reasonable tone-quality, though inferior to the ordinary fingering. It affords the best way of managing the common F′ to E♭ slur and trill (0 –23* 4–––), and the trill is easily turned by adding a finger (5) to give a D fingering already mentioned. Unfortunately the basic fingering of this group of alternatives is sometimes sharp and it must be flattened by adding a finger below such as 0 –23 4–6– (the F′ to E♭ trill still works with 3). The trill can, however, be played with an F′ alternative followed by one of the first group of E♭ alternatives, e.g. 0 1*–3 4–6–, although this method may be criticized on the grounds that the upper note of a trill should, wherever possible, be a normal fingering, both to give the best tonal results and to make the trill easier to negotiate in the stress of performance. As it is quite impossible to trill neatly from F′ to E♭ with normal fingerings for both notes, the execution of this trill and its turn is a problem that every serious recorder-player has to solve for himself.

The E♭ fingerings – 123 45–– and ––23 456– have little or no use. They do, however, complete the pattern of alternatives. The first and biggest group of E♭ fingerings consists of forked flattenings of the note E; the second group are flattenings of F'; the third is a flattening of the normal F♯', and the last of G'. The more remote the original note is, the heavier the forking has to be, and, in consequence, the poorer in quality the alternative and the more difficult to tongue with any degree of attack.

On many renaissance recorders and on some basses, E♭ is fingered 0 1–3 –––– (i.e. normal baroque fingering without 4), or 0 1–3 –5––.

E: The alternative 0 –23 –––– is so well known that it has been taken for the normal fingering of E. The plain-fingered 0 1–– –––– gives a richer note, however, and the fingering is easier to manipulate, particularly in the treble's basic key of G major. The alternative, which is not perfectly in tune on all instruments, is most useful in F' to E slurs and trills (with a 0 –23 45–– D turn), and it also reduces the number of finger movements in slurs from the upper octave such as C' to E. It may also be preferable to the main fingering in the common G' to E slur, but this, together with its use with C, is a matter of personal choice.

The flattening of F♯', already a forked fingering, gives – 123 4––– as a fairly poor quality alternative E: it may be sharp but can easily be flattened by further forking (with further tonal damage). This alternative is of use for an F♯' to E trill as follows: – 123* 4–6–, and as the turn after a G'–F♯' trill (*PB* Ex. 59).

The two usual E fingerings, the plain fingering and the 0 –23 –––– alternative, seem to meet all exigencies, for they are both amenable to intonation shifts on the principles set out in Chapter IV. The plain fingering can be flattened by wide forks such as 0 1–– 4–––, and the alternative may be flattened in the same way with 5 or sharpened by double forks such as 0 –2– 456–.

F': F' alternatives have already been discussed in some detail (p. 62); if they were all counted it would not be surprising to find that there were more than a hundred possible fingerings for F' (allowing half-hole variants and tonally useless alternatives).

The main alternative is – 123 ––––. It tends to sharpness whereas the normal fingering tends to flatness: it is therefore very useful for dynamic purposes, particularly as it is amenable to further sharpening. It is invaluable in passage-work moving from the lower to the upper octave across F′ as it allows time for the thumb to move into place (*PB* Ex. 29): play rapidly C′ F′ C F′ with both ordinary and alternative fingerings to prove the value of the alternative (see also p. 69). This fingering should always be used to play E♯′ partly because it is tonally weaker than the tonic F♯′, and partly because it makes the F♯′ to E♯′ trill easy to negotiate with –123 45–– as the D♯ turn.

The other important alternative, with a surprisingly good tone-quality, is 0 ––3 4––– (sharpish) or 0 ––3 45–– (flattish), with 0 ––3 4–6– as a variant. The latter fingering provides an F′ to E♭ trill already mentioned (under E♭ above).

F′ may be played Ø 1–– 45––, Ø 1–– 4–––, or Ø 1–– –––– according to the amount of thumb opening. This poor-quality alternative has obvious uses in quick jumps or slurs from top F″; octave jumps from E′ to E can be managed in much the same way.

A useless F′ alternative is 0 ––– 4567: but it is interesting because, contrary to expectations, sliding the little finger across on to both its half-holes does not always flatten the note but may cause it to 'break' upwards to between F♯′ and G′, a phenomenon peculiar to the upper octave.

Alternative fingerings in the upper register

The exact position of the thumb makes or mars the production of good notes with alternative fingers in the upper octave. To describe the movement of the thumb in its octaving function it is preferable to use the word 'thumbing', rather than 'pinching', although the latter appears in eighteenth-century tutors. This is partly to discourage the taut and unrelaxed positioning of the thumb implied in the word 'pinching' (some players press so hard against the side of the thumb-hole that they actually bend their thumb-nail), and partly to facilitate the employment of the method of octaving which involves drawing the thumb slightly to one side from the closed position and opening a crevice between the *flesh* of the thumb and the rim of its hole instead of between the *thumb-nail* and the rim as in 'pinching'. This method of octaving requires the minimum of movement from the position in which the thumb-hole is closed, so long as the thumb is always kept fairly upright (see *Introduction to the Recorder*, illustrations on pp. 32, 33, and 22). The thumb movement need be no more than a pivot on the soft flesh of the thumb, a smaller movement than 'pinching'. Compare the ease with which rapid C to C′ jumps can be managed by pivot thumbing with the awkwardness of alternately pinching and closing the thumb-hole. Note that in pivot thumbing the thumb-nail may still be touching the recorder, as it may in the fully closed position, but it touches just outside, not inside, the thumb-hole rim. The disadvantage of this method is that it is less easy to make the minute measurements of thumbing necessary for the perfect production of high notes (see Chapter VI) than it is with the thumb-nail; octave slurs from D to D′ will demonstrate this. The ideal is a combination of the two methods, the thumb-nail being brought into use for greater precision or when accurate thumbing is called for on high notes (*PB* Ex. 35).

Widening the thumb aperture in the upper octave usually causes sharpening of the note: with the thumbed F♯′ (see below) the sharpening is considerable; with A′, however, there is no such effect (a useful phenomenon); and with higher notes such as C′ or C♯′ the sharpening is limited by the breaking of the note when the aperture becomes too wide and ceases to have its octaving effect – within the bounds of safety, however, the sharpening is sufficient to constitute an extremely valuable intonation control.

F♯′: G♭′: The normal fingering – 12– ––––– is a forked flattening of G′: it can itself be flattened by further forking, e.g. –12– 45––, although with too many fingers added it may go down to F′ or, surprisingly, break up to A♭′ (e.g. – 12– 456⁊).

The note may be sharpened (for it often offends by flatness) by double-forking, e.g. – 1–3 4–––, or – 1–3 –5–7: there are many variations.

The fork may be widened to give either sharper or flatter alternatives such as – –23 –––– (sharper), or – –23 4–6(7) (flatter). The even wider fork – –3 456– is a curiosity because it seems uncertain as to whether it wants to play F♯′ or A♭′; its efforts at both are poor.

F♯′ played 0 ––– –––– is probably too powerful in tone-quality for normal use, but 0 1*–– –––– gives a heroic F♯' to E trill on a D major cadence. Virdung (1511) and other renaissance writers, however, give it as normal fingering for F♯', for it is more even in quality with its neighbours on a renaissance wide-bore recorder. Baroque F♯' tends to be flat on renaissance recorders: other possible fingerings are – 1–3 ––––, – 1–3 4–––, or – 1–3 –5––.

F♯' is the first note we have got to which responds to thumbing, though its tone-quality, too, sounds 'pinched'; the thumb aperture should be small: Ø 123 4567. It can be used for an F♯' turn after an A' to G' trill resolving upwards, or in fast slurs with high notes.

G': Like all notes in the middle of the recorder's compass, G' can claim a vast number of fingerings, particularly as – ––– –––– gives a G' (very sharp and coarse). One finger on anywhere gives G' although – 1–– –––– is rather flat, and most combinations of two or three fingers also produce it – a veritable *embarras de richesse*. The only useful one of this huge group of alternatives is – ––3 4–––– which is handy for the common E♭ to G' slur as it makes it a two- instead of a four-finger movement. If normal G' is flat (e.g. with a renaissance recorder), it can be played with 3 instead of 2. This is also useful as a soft G' fingering.

By far the most important G' alternative, and the second in importance of all alternatives, is – 123 4567. It is unfortunately rather coarse in tone-quality and must therefore be treated gently, but it is nevertheless indispensable for the trill A' to G' which is fingered – 123 45*67: it is the best way to play this common trill effectively, although – 123 456*7*, and Ø 123 456*7(*), are possibilities, the latter being useful when – 123 4–67 gives a sharp A'. When this trill, as it often does, commences with an A' appoggiatura, the A' should be thumbed, for with an open thumb-hole it is coarse in quality and sometimes difficult to tongue; the thumb should be moved immediately the trill begins or the G' will be very flat. If the trill finishes with an F' or F♯' turn, the G' before the turn should be the normal fingering, and so, generally, should the G' after the turn. The semiquaver phrase (slurred) A'–G'–A'–G'–A'–G'–F'–G', with the first two G's alternatives and the last two ordinary, should be practised to perfection (see p. 119).

On most recorders the upward slur from G' to A' tends to

'click' with normal fingering, and it is well to become accustomed to use the – 123 4567 G′ alternative for this slur in almost all contexts. The only objection to its use is in slow music where the coarseness of the G′ alternative may become noticeable and spoil a phrase by tonal ugliness or false accentuation: the solution, then, is to tongue on normal fingerings with such subtlety that an impression of slurring is given. The use of the G′ alternative should be almost a reflex action for G′–A′–B′ and G′–A′–B♭′ slurs, fingered – 123 4567, – 123 45– –, Ø 123 –5– –, and – 123 4567, – 123 4–67, Ø 123 4–6– respectively, in each case the thumb being moved into its octaving position during the playing of the A′. In slurred D major runs or scales (which are frequent in recorder music) the G′ alternative should always be used in the upward slur, firstly because it facilitates fingering, the first finger of the left hand being able to stay down, secondly because the dangers of 'clicking' between F♯′ and alternative G′ are far less than between ordinary G′ and A′, and thirdly because the finger movement favours the shape of the run (see under 'Runs', p. 68). In F′ to G′ (alternative) slurs the likelihood of 'clicks' is greater than in F♯′ to G′ (alternative) slurs, and fingering in an F′–G′–A′ slur is therefore a matter of choice, dependent upon the musical demands of the context, upon personal preference, and upon how reluctant and 'clickish' your recorder is when crossing the break – some recorders are able to cross the break with elegant smoothness in which case normal fingerings can always be used instead of the clumsier alternative G′. Rapid F′–G′–A′ slurs can easily be managed, almost without 'clicks', by using an open thumb-hole alternative for F′, e.g. – 123 – – – –, followed by – 123 4567 and – 123 45– –. In slurs down from A′ to G′ use the ordinary G′ and not the alternative fingering (unless the slur goes down and then up again).

The note G′ may be produced by the thumbed alternative Ø 123 456⁊. As the tone-quality of this note is somewhat stifled it is only useful in rapid music, and has to be used with caution. Its value is in the mordent A′–G′–A′, in a G′ turn after a B′ to A′ trill, and in certain slurs and jumps to high notes where neatness at speed can only be managed if the thumb stays still. With or without ⁊ it has to be used for A′–G′ trills if the unthumbed alternative is flat.

A♭′: G♯′: This note has two fingerings, analogous to the main G′ alternative and to the thumbed version, each with comparable tone-qualities. The ordinary fingering is generally given as − −23 456–, or the same with the first finger down (flatter and coarser). It will still strike on most instruments without the second finger down (i.e. − −−3 456–).

The sharpest alternative is − −−3 456–, and it is also the coarsest. Even this, however, is on some instruments not as sharp as it should be, and an accurate A♭′ can only be achieved by octaving the lower register fingering Ø 123 45̶6–, which then becomes the normal fingering. This note is true but comparatively subdued, its tonal kinship being with E♭′ rather than G′. It is easily flattened (little finger shading) or sharpened by slide-fingering.

Players who are able to use both the thumbed and the open A♭′s may exercise their musicianship by suiting the fingering to the context. The easy sharpening of the thumbed A♭′ makes it more suitable for use in quiet passages; while the open A♭′ is used where accent is required. Jumps to the lower register are more easily managed with the thumbed fingering, particularly if the pivot method is used, but here again the context may require the more powerful note.

The A♭′ to G trill is a crux in recorder playing. Here are twelve possibilities to try, not all of them of value:

(i)	−	1 2 3	4*5 6 7 (A♭′ rather flat).
(ii)	−	1*2 3	4*5 6 7 (accurate but difficult).
(iii)	−	1 2 3	4 5 6 7* (A♭′ rather flat).
(iv)	−	1*2 3	4 5 6 7* (accurate but very difficult).
(v)	Ø	1 2 3	4 5 6̶ 7* (G rather sharp).
(vi)	−	− 2*3*	4 5 6 − (like the remainder, 'clickish').
(vii)	−	−−3*	4 5 6 −
(viii)	−	− 2 3*	4*5 6 − (gives an F′turn with the thumb).
(ix)	−	−−3*	4*5 6 −
(x)	−	−−3	4*5 6 −
(xi)	−	−−3	4 5*6 −
(xii)	−	−−3	4 5 6*− (G rather flat).

A choice of evils lies before the recorder-player: the first is to be favoured for very fast trills and the second where intonation becomes a consideration, while the ninth has a pleasing bright neatness admirable in some contexts.

The difficult G♯′ to F♯′ trill may be fingered − −23 4*56–,

– X̶23 4*5*6–, or – 123* 4*56–, all being decidedly 'clickish'. Alternatively it can be played without 'clicks' with the fingering Ø 123 456*7, 6 trilling low as Ø 123 45–7 is a sharp G♯'.

A': By a very short head, the best fingering for A' is the normal Ø 123 45––, but – 123 45–– and – –23 45–– are identical in intonation and only slightly less good in tone-quality (compare the windy undertone produced by each). They are similar enough in quality that, except in slow music, the thumb may be brought into the octaving position during the playing of the note. This characteristic is taken advantage of in the G'–A'–B' and G'–A'–B♭' slur fingerings already referred to, and it should be exploited whenever occasion demands so that the thumb can be carefully placed in the correct octaving position for the following notes (this is yet another good reason for always reading a few notes ahead of the note you are actually playing). Generally speaking the proximity of G's and F♯'s calls for an open A' while higher notes and lower octave notes suggest the normal fingering for A', particularly if pivot thumbing is used for jumps from the low octave.

In the G'–A'–B♭' slur fingering the note A' is produced by the fingering – (or Ø) 123 4–67. This is a weaker fingering as it needs light tonguing and is a little coarse in quality with the thumb off. It is useful in the B♭' to A' trill, although one needs a supple little finger to keep it going long. A sharp B♭' to A' trill is Ø 123 45*––, but it can be brought into tune by trilling with 5 very low. Trilling with 7 is preferable.

Another crux in recorder technique is how to negotiate the common slur F' to A'. The only answer is to use normal fingering, pivot thumbing, and careful tonguing. At first the proposition – 123 –––– to – 123 45–– looks attractive, but in fact it is very 'clickish' and this fingering only comes into its own when applied, trumpet-like, to the slurred fanfare C'–A'–F'–C–A–C–F'–A'–C', or, without upward slurring, to passage-work (see above under the note F'). A slight increase in breath-pressure at the moment of the slur up to A' helps considerably, but it is agility of fingering and thumbing that counts most.

B♭': A♯': This note needs to be played with careful control of breath-pressure as on some recorders it tends to blurt. It may be slightly sharpened by thumb movement or by slide-fingering (to which, as a forked fingering, it responds well), but the thumb should never be removed altogether for although the note does not 'break' it becomes extremely loud and coarse. Ø 123 4–67 gives a slightly sharp Bb′ that is useful before the trill with Ab′ Ø 123 45*6–. This is a more accurate trill than the coarse and sharp – 123 45*6– fingering, which can, however, be moliified by a touch of thumbing. Ø 123 4––7 gives a sharp (or soft) Bb′.

A slur or triil from B′ to A♯′ is taken Ø 123 –56*– with 4 going down for the turn. This A♯′ fingering is tongueable but sharpish. The right hand should adopt the wrist pivot half-hole position for this alternative.

B′: As with all forked fingerings, intonation control is easy to manage, quite apart from the use of the sharpening effect of widening the thumb crevice – at least to the point where tone deteriorates or the note 'breaks'.

The only alternative worth mentioning is Ø 12– 45–– which can be used as a mordent or turn after C♯′. In using it breath-pressure must drop slightly or it will strike E′, for which it is, of course, the normal fingering.

C′: When buying a recorder, one of the things that should be tested is the accuracy of the note C′ with ordinary fingering and the widest practical thumbing aperture, for C′ is a dictator among the notes of a recorder as it has virtually no alternatives and cannot therefore easily be sharpened. Slide-fingering with 3 is to be avoided, but it is effective with 1. Flattening is of course straightforward, either by ordinary or by little-finger shading. The effect of the flattening Ø 123 ––6– is just about slight enough to make a one-finger movement of the C′ to Bb′ trill: Ø 123 4*–6–.

Mordents or trill turns after C♯′ may be fingered Ø 12– 4–6–, but this C′ (or B♯′) cannot be tongued with security.

C♯′: Db′: This note requires at least two-thirds coverage of the thumb-hole, preferably an aperture equivalent to one-tenth or less of the area of the thumb-hole (though optimum thumbing apertures for high notes will vary from one recorder to another – the fractions given in this book are only guidelines). Rapid repetitions require light, precise tonguing and close thumbing. There is enough latitude of thumb movement to effect some sharpening (which is often needed as C♯′ is flat on many recorders), but as this note often tries to clear its throat before speaking it is wise to begin with close thumbing and the moment the note speaks to slip the thumb across to sharpen it – it should all be done so quickly that nobody notices. It is amenable to slide-fingering with 1, and shading with 7.

A useful C♯′ alternative in slurs (it is difficult to tongue) is Ø 12– –5– –. This is handy for the common A′–B′–C♯′–D′ slur (or its reverse). As is to be expected of a wider fork this fingering for C♯′ is slightly sharper than the normal fingering.

Ø 123 4567 is a sharp C♯′ fingering useful in slurs down from E′ and D♯′. It is impossible to slur from D♯′ to the slow-speaking normal C♯′ without a 'click', so trills should be done with this fingering, trilling with 3 and 7 together, or, better, with 2.

We have now reached another turning point in the mechanics of the recorder, for contrary to expectations, adding a finger below C♯′ results in an upward break instead of flattening. This phenomenon is the basis of high note fingerings described in the next chapter. It does, however, give the C♯′ fingering Ø 123 –567. On renaissance wide-bore recorders and with some basses, normal C♯′ can be so difficult to tongue that it is too unreliable for general use, whereas Ø 123–567 gives a sweet and responsive C♯′.

Extended alternative fingerings

Lest it be thought that the foregoing is a complete review of alternative fingerings up to C♯′ I would refer the reader to Michael Vetter's book *Il flauto dolce ed acerbo* (instructions and exercises for players of new recorder music) published by Moeck, Celle, 1969, but currently out of print. This lists several hundred fingerings for standard pitches of the open register, i.e. with no bell-key or the bell-key open, for chords, and for notes incorporating white noise ('rustling noises of air and harmonics'), and then repeats the process for the fully closed register, and for the covered register, i.e. with the bell covered but not airtight. Other lists include thumbed harmonics in the lower octave (p. 60), and similar very thin sounds in the upper octave.

To take an example, reference is made on p. 71 above to playing bottom F *ppp* and in tune by slide-fingering 7. In practice this can be achieved by slide-fingering, or rather just leaking, *any* finger or the thumb, and each thin quiet note produced has a slightly different quality. A contemporary composer may wish to use a particular sound, probably electronically amplified.

Another review of fingerings, much shorter and more critical, appears on pp. 81–106 of Daniel Waitzman's *The Art of Playing the Recorder*. This includes bell-key fingerings, but is nevertheless a useful commentary for players of keyless treble recorders. The treble recorder, unlike recorders of other sizes, can be shaded or even covered by lowering it upon the upper leg (see pp. 70 and 89). If it is fully closed (airtight) with all fingers and thumb on, the treble recorder produces a cavernous low Bb. With other fingerings the closed, covered, or shaded bell produces a wide range of interesting notes, increasing the recorder's tonal range for use in contemporary music. There is then, however, the problem of taking account, while playing, of the fingering required by the composer and indicated in the score, or of learning the piece by heart with the special fingerings.

To attempt to play avant-garde music certainly increases one's knowledge of the recorder's ultimate capacities. Berio's *Gesti*, for example, dissociates fingering from articulation so that a percussive noise (based on a Telemann duet) is produced by the fingers alone while the breath input is itself not synchronized with the fingering. Such a piece needs to be prefaced by an explanation of the signs used in the score – but it may be said this is no more than was the case with eighteenth-century French music. Whether players will wish to master this new vocabulary will depend on their adventurousness, their open-mindedness and their judgment as to whether the reward in musical satisfaction is worth the considerable effort of so greatly extending their technique.

VI
HIGH NOTES

Readers who have developed the good habit of following this book recorder in hand should now shut windows and warn neighbours, for it is inevitable in introducing oneself to the highest notes on the recorder that loud and unpleasant noises are perpetrated before they become softer and sweeter through familiarity and a growing command of tonguing and thumbing.

Tonguing and thumbing – these are the clues to success in the highest register. Look back to Chapter III and practise again some of the exercises suggested there until you are sure you have learned the knack of altering the strength of articulation in the tonguing without changing the breath-pressure on the note itself. To obtain high notes cleanly, use a precise tonguing, such as 't' or 'd'. It is precision of tonguing that matters, i.e. the speed and pointedness of the tongue stroke on to the teeth-ridge, not its strength (i.e. how firmly the tongue is pressed on to the teeth-ridge and air-pressure built up before the note starts). Nevertheless high notes will respond to variations in the strength of tonguing. Practice D′ with 't' tonguing, varying from a strong 't' down to the lightest 't' at which the note will strike. You will find that it responds better to the lighter tonguings: the strong tonguing will give a less clean articulation, for the D′ will be mingled with elements of the next harmonic, A″. If you tongue more strongly, you will in fact strike A″ or thereabouts. It is a very common fault at the intermediate stage of recorder playing to tongue high notes too strongly. Having felt the range of tonguings within which D′ can adequately be articulated, construct for yourself a gradation of light, medium, and strong tonguings for D′. Now on each of these play the note in turn *f*, *mp*, and *pp* (i.e. nine variants) keeping within the range at which the note, once struck, can be played with an acceptable tone-quality at different volumes (ignore intonation changes for the time being).

You should next see what happens when you articulate repeated D′s with double-tonguing. If you try your soft 'diri-diri's, you will have difficulty in getting the note to start cleanly with 'r'. This is too imprecise an attack for high notes. 'l' is not much better. High notes will respond to 'k', within the double-tonguing 'tee-kee tee-kee', although they come more easily on the 't' stroke. Quite apart from fingering problems, this is why fast passage-work at the extreme of the recorder's compass is so difficult, unless the voicing and construction of your recorder especially favours high notes. Generally, try and manage high notes, even at some speed, with single-tonguings (*PB* Ex. 46). The shortness of the tongue movement using 'tee's' and 'dee's' helps.

Consider next the position of the tongue in the mouth between the tonguing strokes. It should be high, in order to make your column of breath flow through the mouth at a higher pressure, but using no more air than for lower notes. This point is explained more fully on pp. 103–4. The vital thing to keep in mind is that as you go higher on the recorder you change from 'er' for the middle notes to 'ee' for the high notes. You 'smile your way up' to the high notes.

Now turn your attention to the thumb, still practising on the note D'. As 'pinched' thumbing is more reliable than 'pivot' thumbing for high notes, first cut the nail fairly short. Place the thumb into its hole, and, in as relaxed a manner as possible, move the thumb about the hole, gently feeling its edges. Try pivoting on the thumb-nail so that the fleshy side of the thumb is lifted slightly away from the far side of the thumb-hole. All the while the thumb should be absolutely relaxed, and should touch the hole as if afraid to damage it. It should touch its hole so lightly that, if someone gave it a gentle flick while you are playing, it would be pushed away. Of course, if the thumb-hole is damaged or worn down by harsh thumbing it should be re-bushed, for it is not easy to produce good high notes with a worn thumb-hole.

Apart from controlled tonguing and sensitive and relaxed thumbing there is another prerequisite to the successful playing of high notes. Drops of moisture in the tone-producing areas have a pernicious effect on high notes, and as air is pushed more rapidly through the windway in producing high notes, conditions are conducive to wetness. It is therefore essential that for playing a piece containing high notes the recorder should start clean, warm, and dry.

In fingering all notes of the recorder, and particularly high notes, all the fingers, and indeed the whole hand, should be relaxed. Each finger should rest firmly but gently on its hole, lightly enough for an infinitesimal movement to be sufficient to make the difference between closing and just not quite closing the hole. Let us now proceed to the fingerings for each note.

D′: This is not always an easy note to produce. Its normal fingering, Ø 12– –––– requires no more than medium tonguing, otherwise it will overblow. It is often flat, but may be sharpened by widening the thumb aperture immediately after striking; this sharpening device may be used on all high notes up to half opening the thumb-hole. D′ will only strike with a fairly small aperture – never less than eight-tenths closed, but on the other hand too close thumbing will make the note harsh and tight pinching therefore spoils it. It is probably better to slide-finger 1 for soft D′s. Shade with 7 for loud D′s.

A wheezy alternative for D′ is Ø 12*3* 456*7*, useful mainly for the trill after E♭′. *3* should be about three-quarters shut controlling the intonation, and the trill should be executed with the third finger of the right hand (6). This fingering is also used with E′ to D′ trills with either 6*7* or 2 doing the trilling (*PB* Ex. 60). The alternative D′ is very useful for turns after trills involving E′. A coarser variant of it is to cover 7 entirely and make *3* only half shut; in this case the trilling finger for the E♭′ to D′ trill is 7, and for the E′ to D′ trill is 2.

In the previous chapter we saw how on a wide-bore recorder, reluctance to speak C♯′ with Ø 12– 4––– could be overcome by using Ø 123 –567. Similarly, as the end of the compass of a wide-bore recorder is reached, problems may be encountered in getting D′ to speak. The apparently perverse fingering Ø 12– 4567 may in fact be the solution. Some renaissance recorders, however, are happier with only one added finger, e.g. Ø 12– –––7. In this respect recorders react in the same way as reed-cap instruments – they seem to like 'stabilizing' right-hand fingers added for notes made with left-hand fingers. The same may apply to bass recorders, although a bass recorder tends to be a law unto itself in fingering high notes. D′ may tongue adequately on an F-bass recorder with normal fingering, but you may find that it becomes more responsive to a range of tonguings if stabilizing fingers are added. The same may apply to C♯′, and even to C′.

E♭′: D♯′: The normal fingering, Ø 12– 456–, gives a clear and lovely note on most instruments, even on basses and wide-bore instruments that are capricious on D' and C♯'. The thumbing is not critical – the note can be induced to strike with only three-tenths of the thumb-hole closed: the optimum thumbing is nine-tenths shut. For pianissimo playing the third finger of the right hand may be brought back to its half-hole: the note can be flattened by moving the little finger across to its half-hole. These could be your normal fingerings if you need to correct the tuning.

Alternatives for E♭' are Ø 12– ––6̸7, Ø 12– ––67 and Ø 12– –––7, all of which can be used for trills to D', the last, if it works, being the least 'clickish'. Another is Ø 1–– 4567, though this may be so sharp as to be a flat E'.

E' is another easy high note with Ø 12– 45–– as its standard fingering. It responds best to firm, precise tonguing, and strikes at seven-tenths thumbing, the optimum being just over eight-tenths, a slightly wider aperture than the best D'.

An alternative available on some instruments for slurs after D', and trills E' to D' (using normal fingering for the first E') is Ø 1–– –––– With this fingering it is usually necessary for the thumbing to be as close and tight as possible, for the trilling finger (2) to trill low, and for the breath-pressure to be no more than medium, otherwise – strange phenomenon! – the fingering may give F‴ instead of E'. A fast D'–E'–F‴ slur is best played with this fingering for E' as the 'click' between E' and F‴ is much less than the 'click' across the powerful register break between D' and E' with normal fingerings.

E' is sharp on some instruments and may then need 7̸ to be in tune.

E' can be obtained with good tone and quick tonguing response by flattening F‴ with 6̸7 or 67. Over-tonguing must be avoided or the fingering will strike G‴. This fingering can easily be sharpened or flattened so it is useful for obtaining dynamic gradations. Daniel Waitzman also advocates it for slurred leaps from below (try from B').

On basses and renaissance instruments the starting point for E' is always the 'two left, two right' normal fingering, although the two right-hand covered holes could easily be 56 instead of 45. If E' is flat, leaking 1 or 2 may be the solution, or even taking 2 off altogether, although it is disconcerting to have to finger normal F‴ to play E'.

F‴ needs considerable care. Its normal fingering is Ø 1–– 45––

and the optimum thumbing is nine-tenths. With this thumbing F''' may be played fairly softly. More than nine-tenths thumbing makes a harsh F''', so it is a fallacy that for top F''' the tightest possible thumbing should be used. With less than seven-tenths thumbing, F''' will not strike. Experiment with F'''s of different volume (although *mp* is about as soft as one can safely go), and gradually the note will present fewer terrors – and it is fear of the note that causes unrelaxed thumbing, over-tonguing, and disaster.

F''' is often flat, but fortunately it is amenable (more so than E') to sharpening by widening the thumb aperture. It can also be sharpened by extremely cautious slide-fingering of 1, or by fingering with 5 leaking. If it is sharp, add 6. This is in fact a more stable F''' and tongues more easily.

Repetitions of F''' are difficult because the note is very slow-speaking. A possible solution is to form the thumb aperture on the side of the thumb-hole farthest from the mouth; this may be done by allowing the thumb-nail to rest so closely against the near side of the hole (as if it were taking some of the weight of the instrument) that there is little or no thumbing aperture left. Then bend the thumb slightly back towards the mouth so that it is eased away from the far side of the hole. This method of creating the thumb aperture ('double thumbing') results in quicker-speaking high notes, but it is harder to control the extent of the opening.

Always remember to 'smile at' top F''' for it responds best to a thin high-pressure air-stream. Whisper 'tee' into the instrument, and with accurate relaxed thumbing it will not fail you. Provided the tonguing remains precise, it is surprising how lightly one can then tongue and still strike the note.

On some instruments, F''' is not an easy note to sustain with good tone-quality, as the undertone can be threatening. Thumbing has to be very accurate and breath-pressure very steady, maintaining 'eee', or the tone-quality will diminish. If its quality is generally not good, breath-vibrato may help to sweeten it.

Wide-bore instruments are likely to have no F''', except by rather unsatisfactory slurring up from E' (but see p. 91).

Basses may produce F''' with normal fingering, but, as always, it is worth experimenting to find the best fingering. This may turn out to be the first finger of the left hand plus almost any grouping of fingers in the right. Or it may be a flattening of normal G″, such as Ø 123 4567 with 6 leaking. It is worth lowering the position of the speaker hole (i.e. octaving aperture) further down the bore than 'double thumbing' brings about, by covering the thumb-hole completely, or almost closing it with the pinched thumb, and using a leaking 1 as the speaker hole (if you put the little finger of the left hand on the side of the instrument, this facilitates control of 1). This is a valuable device for high notes as it can improve their intonation and tone-quality. On the other hand it is no easy matter when playing a group of notes to make the thumb/first finger change, and to get the first finger leaking by just the right amount. It takes a great deal of practice. But the result could be the ability to play a sweet-toned, accurate, and quiet top F‴ instead of an unreliable and coarse loud note.

F♯″: G♭″: This is the bugbear of recorder players. F♯″ is obtainable by a number of methods, but none of them is entirely satisfactory. The best known is Ø --- 45--; this can seldom be coaxed to speak without sliding up to it from F″ or E′. Alternatively, F♯′ will play on Ø --- ---- in the slur from D′. Slight shading with the first finger assists the note, particularly if it has to be sustained: thumbing is as for F″. This note nearly always suffers from flatness. Some recorders offer an F♯″ with Ø 1-- 4---, with 1 leaking, but this requires perfect thumbing and tonguing.

The other F♯″ is generally too sharp: it is fingered Ø 123 4567, being easier to strike and pleasanter in tone-quality with 5 lifted fractionally from its hole; or leaking 6 instead may make it flatter. A further way of flattening it is to use the first or even the second finger as the octaving hole. This makes the fingering 0 123 4567 with 2 and 5 (or 6) leaking. If such a fingering for F♯″ is played softly it may be in tune.

A third method of getting F♯″ is to use the flattened G″ just described and flatten it still further by shading the hole in the foot of the recorder against the upper leg, a device available to treble players only, provided that they are sitting down. But in fact, by lowering the instrument so that all the bell opening is covered, a perfectly good F♯″ will play on normal G″ fingering. It does not matter that the covering is not airtight against the material of one's clothing. If you see an F♯″ looming up (e.g. in Telemann duets), you should bend forward very slightly so that the end of the recorder is close to the leg. The final little plunge can then be made inconspicuously actually at the F♯″. You need to be cautious about this movement, as if you are too energetic you may damage your teeth. But your lips, which should anyhow already be in a slightly stretched position for playing high notes, can tauten up to cushion the slight thrust of the recorder end into your leg without putting your front teeth at risk. You have to be quick to get back into the uncovered position to play the next note. We shall see that other high notes are more easily playable with bell-covered fingerings, so it is a technique well worth acquiring. It will effectively slur down the octave to F♯′ even with the end still covered (*PB* Exxs. 48 and 49). It should not be used for the high F♯″s at the end of the Herbert Murrill sonata (*PB* Ex. 47), because of the faster speed of the piece and because the composer knew that high F♯″s were only available slurred up to from beneath by a soloist standing up.

Bass players will usually find an F♯″ fingering by getting to know the idiosyncracies of their instrument. It will probably be a flattening of normal G″, possibly all fingers on with a leaking fingering somewhere. If normal G″ happens to be flat, so much the better, for this increases the chance of finding an F♯″, and normal G″ fingerings can easily be sharpened.

G″: This note is gratifyingly easy to play with fingerings based on Ø 1–3 4–6–. Flatter variants are Ø 1–3 4––7 or Ø 1–3 4–67; sharper Ø –23 4–67. Recorders differ as to which G″ fingerings they prefer – each player must experiment with his own instrument and find what suits it best. The tonguing and thumbing are the same as for E′, that is to say, there is a fair amount of latitude.

Slurs involving G″ are difficult because it is too persistent to slur down to F″ without a considerable 'click' across the register break. There is no way out except to slur 'portamento'. The slur to F♯″ is easier with the Ø 123 4567 fingering, but the semitone is rather a slender one, and where this slur appears a resourceful player would choose to use one of the sharper fingerings for G″, giving a much more convincing G major scale, despite the inevitable click between F♯″ and E′.

If F♯″ is played with normal G″ bell-covered, it is useful to have a bell-covered fingering for G″ itself as the two notes are often in proximity. Bell-covering in this register lowers a note by a semitone, so bell-covered G″ is based on normal A♭″. Oddly, in the register to which normal E′ and F″ belong, bell-covering raises a note by a semitone, so providing an alternative bell-covered F″ fingering with normal E′ Ø 12– 45––. Bell-covering works just as well on a wide-bore renaissance treble recorder; my Moeck recorder will give F″, otherwise missing, with bell-covered Ø 1–3 45–7, and an E′ with bell-covered Ø 1–3 4567, although a 'normal' E′ is available with Ø 1–– –56–, or even just with 1. The problem with bell-covered notes is that it is very confusing to play a normal fingering and then get a note a semitone different. Only considerable practice will enable these notes to be used with confidence in actual playing. Certainly bell-covered F♯″ should be mastered because of its accuracy and good quality. Bell-covering for A″ and B♭″ provides alternatives to notes which are otherwise tonally barely usable, so these bell-covered notes come in useful if, for example, one wants to play Bach's solo cello suites on the recorder (very difficult but infinitely rewarding).

A♭″: G♯″: This note is fingered Ø 1–3 –––7, with the standard pattern of variants, and is best tongued firmly and precisely with nine-tenths thumbing.

A″: The fingering is Ø –2– ––––, and needs strong tonguing and nine-tenths thumbing. This note appears at the end of the first movement of a sonatina for treble recorder and piano by P. Glanville-Hicks. A″ is flat and coarse on many recorders but can be sharpened with slide-fingering. The bell-covered Ø –23 –56– is a much better note.

B♭″: A noise somewhere near B♭″, but too sharp, may be obtained by forcing a hurricane into the recorder with the fingering Ø 123 4567 with 3 leaking. A perfectly adequate treble B♭″ is obtainable, however, by bell-covering normal B″.

B″: This is quite a good note, though always loud, with the fingering Ø 12– 45––, strong tonguing (so that it does not play E′), and nine-tenths thumbing.

C″: The normal fingering is Ø 1–– 4–––, but there are flatter variants. The thumbing is as for A″ or B″ and the tonguing even stronger: a sudden-jerked 'whoot' (i.e. no tonguing) may give the best results, but it is difficult to get there quickly after a preceding tongued note. Breath-pressure must be very high. This note was used by Telemann in a recorder sonata.

A bass recorder may without offence to the ear be able to play a complete third octave up to F''', depending on its bore and voicing. Dr. Dolmetsch kindly gave me the fingerings he uses for the upper third octave, which are as follows:

C♯″   Ø 1̸–– 4̸–––
D″   Ø 1–3 –56–
E♭″   Ø ––– –56–
E″   Ø –2– 4567
F'''   Ø –2– –56–

Reverting to the lower third octave, my Stieber great bass (in C) plays C♯″ with Ø 123 4–67, D″ with Ø 1–3 4–6– (normal fingering but quality is improved with 1 leaking for this and all higher notes), E♭″ Ø 12– ––6–, E″ Ø 123 –567 with 3 leaking, F″ Ø 12– 45–7 and F♯″ Ø 12– 4––– with 2 leaking. Experience with other bass recorders suggests that these fingering are reasonable starting-points from which to discover third-octave notes on basses generally. There are higher notes, but those mentioned are all, if played accurately, good enough for serious musical purposes if one can remember the fingerings in time. To play successfully in this third octave demands mastery of shading and leaking-finger techniques, or of other methods of partly covering the finger-holes by the required amounts.

Finding very high notes on recorders is an amusing game to play, but to use these notes in earnest is a different matter. The last two pages of Daniel Waitzman's book take the treble recorder up to B♭″ in the fourth octave, but the author adds, however, that these notes are 'almost impossible to obtain'. Four and a half centuries earlier, Silvestro Ganassi's Chapter 4, 'The art of producing seven more notes on the recorder', makes equally fascinating reading.

VII

VOLUME (DYNAMICS)

In his book *Woodwind Instruments and their History*, Anthony Baines writes, 'Should the treble recorder prove too soft for a modern festival orchestra, then let somebody remodel it to be louder, as has been done with every other woodwind instrument in the course of the last one hundred and fifty years' (p. 75). To say this is to misunderstand the recorder revival, for the recorder is mainly a historical instrument designed to re-create the sounds and style of the music written for it in days before large orchestras were common. Admittedly its incisive and melancholy timbre has appealed to some modern composers of chamber music, but with its volume amplified by acoustical or electrical means (neither method being difficult to put into practice), or simply by plurality of players, it is doubtful whether it would be much of an asset to the standard range of modern orchestral tone. Moreover, any system of making the recorder louder or its working more efficient detracts from its historical genuineness. The recorder is by nature 'une flûte douce'.

The volume of a recorder varies with its voicing, its bore, and the material it is made of (see Chapter I). Moreover, different-sized recorders have different loudnesses – sopranino and descant recorders are louder than trebles, tenors, and basses. But even the lower instruments can be made to penetrate if they are played with a full round tone which gives the impression of loudness because of its clarity and its individuality: a tenor sounding well at its optimum breath-pressure, at which all the breath put into it is translated into tone, sounds louder than if it is blown harder to give a more forceful but less pure note.

Generally speaking, however, the simple rule for a recorder player to make more noise is to put more air into the instrument. It is the greater *amount* of air that creates more sound, not the *velocity* (cf. amps and volts in electricity, or a broad slow-flowing deep river to a rushing narrow stream). Thus high notes can be played softly with a thin high-velocity air-stream, and low notes loudly with big slow-moving exhalation. Especially with a baroque-voiced recorder a substantial range of volume can be achieved without materially affecting intonation. It is a fortunate phenomenon that a note of constant frequency appears to become flatter as it becomes louder, an effect that can be demonstrated with a radio tuning signal. Consequently, if a note increases in pitch and volume at the same time, its sharpening appears less than it actually is: in fact if the actual sharpening is slight the increase in volume may mask it altogether, and the note gives the impression of being constant in pitch. A solo wind-player is thus licensed to roam between certain limits of breath-pressure, and he need only begin to worry when he is noticeably becoming sharp or flat. He is furthermore aided by the accommodating nature of his listeners who, if he is carrying them with him, may accept an A almost a quartertone out of tune if the melody and harmony seem to require an A and the player plays his note convincingly enough. A recorder-player should use volume variations fearlessly to express the melody he is playing: he must listen that he does not go obviously out of tune, but if he gets worried about slight fluctuations in intonation, his audience will do so too, and both his attention and theirs will stray from the music itself and its meaning.

The demands of musical expression cannot be entirely satisfied in the volume range contained within the limits just described; how otherwise does one respond to the thirteen dynamic levels asked for in Vivaldi's music, from pp to ff? A player must therefore be free to blow at any breath-pressure he wishes, so long as good tone is maintained. In doing so he makes for himself an intonation problem, and must apply one of the intonation control techniques mentioned in Chapter IV. In a ff passage, therefore, every note is shaded, no finger rising more than a quarter of an inch from the instrument, with the lowest fingers actually covering their holes for most notes. In a pp passage the unused fingers are lifted clear of the instrument, and most of the plain-fingered notes will be played with alternative fingerings, while notes normally forked will have the lowest component of the fork slid to one side or lifted altogether. It is excellent practice to play each note on the instrument ff, mf, and pp without changing pitch. The note C, for example, is heavily

shaded for playing with maximum breath-pressure, and, as it is
not amenable to slide-fingering, re-fingered for minimum
breath-pressure, thus:

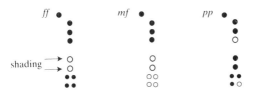

The quality of the refingered *pp* C is not so good as normal C,
but this weakness is masked by the low level of volume.

An even more valuable exercise is to do a gradual crescendo
from *pp* to *ff* on each note of the recorder. A good note to start on
is G'; begin it with no fingers on at all, then as breath-pressure
increases lower the second finger of the left hand to get normal
fingering at *mp*, next begin to shade with the first finger, and
finally add the little finger and the third finger of the right hand
to their holes, with a bit of extra shading with the third finger of
the left if the note has not by then broken: alternatively all the
shading can be done by the first finger of the left hand if it is
moved with extreme care. Similarly practise a decrescendo from
ff to *niente*. When such dynamics are called for in recorder music
it is best to get at once on to a forked fingering with its greater
intonation control, and even the common diminuendo sign is a
warning to begin the note with some shading. A difficult effect to
learn is the sudden diminuendo following a rinforzando; only
careful practice can synchronize the finger movement with the
drop in breath-pressure. Extremes of volume variation should
be mastered if a player is to feel confident in applying the more
normal fluctuations which must be used to give even the
simplest phrase its rise and fall. He should feel that each note is
a being possessed of plasticity in pitch and intensity.

Although it is something of a trick to impress upon a sceptic
that the recorder does have a wide dynamic range, it is also a
valuable exercise to play a two-octave F major scale at your
recorder's extreme limits of loudness and softness. For bottom F
played loudly you will need to shade the bell; also for bottom G,

though you can now bring in little-finger half-hole shading, still shading the bell. For A, take the little finger across to cover both half-holes, and shade 6 as well. For Bb, 5 will need to exert heavy shading by arching its hole, touching the far side. C will have 67 down, and 4 and 5 shading. By the time you reach F′ you will have 567 down, with heavy shading on 3 and 4 and some on 1. A′ will be shaded with 7 down, and Bb with 7 down and 5 shading. Do not take 7 on to both half-holes in the upper octave as this may cause register change, but shade heavily with the others. Top F‴ will have 7 down, and 2 shading to the point of touching the far side of its hole. For the *ppp* scale the bottom octave should be thumbed 'harmonics' (see p. 60). G′ will need to be a thumbed upper-register fingering, e.g. Ø 123 456–, A′ Ø 123 4–67, and Bb′ Ø 123 4–6–. From C″ upwards control is by slide-fingering 1. All your skill will be needed to get F″ to strike softly. It may help if the tiny needle of air-stream (a precise but whispered 'tee') is directed slightly on to the roof of the windway channel. Pay no regard to tone-quality in this exercise: extreme dynamic variation is the only object.

Although finger-movements are the main factor in volume/intonation control on the recorder (unlike on most other wind instruments where the lips can compensate for changes), methods of blowing – quite apart from variation in the amount of air input – can give the impression of volume variations. A note can be given prominence, for example, by playing it with vibrato, and, when a note is being played loudly, vibrato has the double effect of improving its tone-quality and of preventing it from crossing the break upwards.

An effect of variation in dynamics can be achieved by altering the attack accorded to a note (i.e. tonguing), and its duration. This method is used within its limits to shape a phrase, the more prominent notes being played with stronger tonguing and held on longer; or it can be used on a larger scale when whole passages are to be contrasted by the use of 'terraced' dynamics. When extremes of loudness and softness are sought, however, care must be taken not to injure rhythm, which is itself established mainly by tonguing and the comparative durations of notes (see p. 38). If, for example, three crotchets on the same note in a minuet bar are played with equally strong tonguing and are equally long (each note almost being slurred on to the next) the effect is of forceful, continual sound, and so of loudness, but the bar does not constitute a unit of rhythm as no one note is stressed more than another. Slight reductions in the attack on the second and third notes and in their duration (the third note being fractionally shorter and less strongly tongued than the second) provide the differentiation between the three notes necessary to define rhythm, while maintaining conditions in which nearly all the bar is filled with sound. At the other extreme, if an effect of quietness is required the first note may be played with light tonguing and a length equal, perhaps, to a quarter of the duration of the beat, while the second and third notes are played even shorter and slacker, the last just being touched as a pin-point of sound: silence far exceeds sound in such a bar yet the forward movement of rhythm is still apparent. Recorder-players must exploit such methods of obtaining gradations of volume as the natural volume range of their instrument is less than that of most other instruments. The possible volume variations in a bar where there are notes off the beats are legion, although in fast passages it is often sufficient to alter volume only at the on-beat notes, or even the first note of the bar. Experiment with a passage-work echo by shortening only the first of four semiquavers to get some volume reduction: then for greater contrast shorten all the notes in the echo, possibly keeping strong tonguing to achieve the dampened effect mentioned in Chapter III (p. 40).

When a number of people play together, small volume variations can be magnified into big ones, and the method of obtaining dynamics by altering the length of notes becomes more useful still, particularly as the breath-pressure method used outside narrow limits would only add to the intonation problems that already exist in consort playing. Composers of multi-part consorts such as Gabrieli and Gibbons, build in their own dynamic variety by restricting sections of a piece to two or three parts. Changed dynamics in repeats can easily be managed by having fewer players to a part, – perhaps only one to a part, which would facilitate ornamenting the repeated sections. When a recorder is playing with strings or keyboard, the wider volume range of these other instruments can, by sensitive playing, assist the less richly endowed recorder. Subtle changes in speed and mood in consort playing can give the impression of more volume; if the music is exciting and intense it sounds louder, if it is calm and relaxed it sounds softer.

The recorder-player should make the most of his instrument by choosing a position where the acoustics of the room will favour him most, e.g. in a corner which reflects sound. He should not put a screen immediately between himself and his audience in the shape of a sound-absorbing sonata copy perched high on a music stand.

Volume devices

Muting. It is possible to mute a recorder for quiet practice by covering all but the uppermost part of the 'window' with a piece of felt held in place with an elastic band. Another method is to put a piece of paper half in the voicing end of the windway, but this is a messy business when the paper gets wet.

Sound-projection. Volume can be concentrated into an area of a concert room by fitting projections above the walls of the 'window' which throw sound forward instead of allowing it to disperse sideways. This can be done by using Plasticine 'wings' (see p. 63), or with a special plastic attachment shaped like a wheelbarrow without a bottom which clips over the 'window' (obtainable from Dolmetsch). Secondary effects of these devices are that the pitch of the recorder is flattened, so encouraging compensatory higher breath-pressures, and that the 'breaking' point of notes, particularly those in the lower octave, is reached at a considerably higher breath-pressure. All this makes it possible to play the recorder louder, but not without making tone rather rasping.

Bell key. This is an open-standing key operated by the little finger of the right hand to close completely the hole in the foot-joint at the outlet of the bore. The technique for playing the bell-keyed treble recorder is fully described by Daniel Waitzman in *The Art of Playing the Recorder*. Ideally a recorder with a bell key is bored, voiced, and tuned with the bell key in mind. Its compass is greater, and an abundance of fingerings is available to conquer register breaks, afford variety of tone, and improve dynamic resources.

'Echo' key. This consists of a tuning hole (see p. 63), bored so that its opening is in the base of the 'beak' of the recorder, and covered with a closed-standing key operated with the chin (a slight lifting of the player's head does the trick). When this hole is open, the pitch of the instrument is sharpened, or, conversely,

it becomes possible to play much more quietly without flattening. In terms of pitch, the 'echo' key causes a sharpening of about a quartertone. As, with care, this key can be positioned between open and shut, it affords excellent volume and intonation control, making shading techniques look primitive. But then perhaps the recorder itself is somewhat of a primitive instrument. It was said of a certain harspichord maker who continually brought out improved instruments, that if he had gone on long enough he would have invented the piano. Are these unauthenticated sophistications in recorder-making a step towards the same madness? It is a point worth debating.

VIII

TONE

It seems paradoxical to suggest that the simple recorder is a more contrived instrument than the modern flute with its accumulation of keys, but this is true as far as that most basic of functions is concerned – tone production. The recorder-player has sound automatically made out of his blowing by the unalterable relationship between the shape of the windway which moulds to its own cross-section the column of air thrust into it, and of the 'edge' or 'lip' of the instrument which divides the air-stream and forms regularly alternating eddies which produce the vibrations of sound. The flute-player, in common with players of most other orchestral instruments, is able, within limits, to control tone-quality, but he has first to learn difficult techniques: the recorder-player who, like the pianist, produces sound remotely, has the initial advantage of having no such techniques to master, but the ultimate disadvantage of not having much tonal variety at his disposal. He should, therefore, feel himself all the more obliged to study his instrument in order to produce the best tone it is capable of.

Achievement of optimum tone-quality

A recorder-player should know enough about the construction of recorders and the types of tone-quality associated with different shapes of bore (see Chapter I) to judge what tone-quality the maker of his particular instrument had in mind for it to produce. A recorder with a sharply tapered bore cannot be made to lose its reedy quality for the purer, clearer sound of a more cylindrically bored instrument. He should, moreover, not mistake poverty of tone due to some fault caused by bad construction or poor maintenance of the instrument for an inherent characteristic. A breathy or edgy tone may well be the result of an instrument being in need of revoicing or re-bushing, although recorders have on occasion been sent back to the maker for revoicing when all they have needed is a good clean.

The finish of the recorder bore should be such that it encourages the gradual formation of a thin and even film of moisture during play. If this forms too profusely it makes droplets which run down the thumb-hole into the thumb-nail in a most uncomfortable way, or, even worse, block the windway. The recorder should therefore be warmed by a *dry* heat, preferably in the hands or a pocket or against a hot-water bottle at body temperature – more may damage the instrument (fires or radiators are generally too vigorous for recorder warming). A recorder gives its best tone not when it is absolutely dry but when any unevenness in the resonance of the bore of the instrument is smoothed over by a thin film of moisture. To quote Bacon, 'a pipe a little moistened on the inside, but yet so as there be no drops left, maketh a more solemn sound, than if the pipe were dry, but yet with a sweet degree of sibilation or purling. The cause is, for that all things porous being superficially wet, and, as it were, between dry and wet, become a little more even and smooth, and if the body that createth the sound be clean and smooth, it maketh it sweeter.' (*Natural History*, cent. III §229–30, adapted.)

The best made and the best kept recorder, played in a dry, warm room, needs skill and understanding before it will produce perfect tone throughout its range. The player must become aware which notes are weakest and, by concentrated experimentation and listening, should determine the exact breath-pressure required for the best possible sound of each note. Then should follow slow scales and pieces to accustom him to these tiny fluctuations of breath-pressure until they become reflexes conditioned by each particular instrument that is played upon. He will find that some notes are capable of producing stronger tone than others (e.g. G, G', and F'), that certain notes are by nature coarse (e.g. Ab' without thumb, and Bb') and some sweet (e.g. Eb'), and that cross-fingered notes are poorer and fluffier proportionately to the extent of the cross-fingering. In fact, before long the player will find that every note has its own personality. This may be considered one of the charms of the recorder – a wayward inequality of tone that it shares with a good counter-tenor voice, and which seems peculiarly suited to old music. Nevertheless, the recorder-player must cultivate equality of tone as a prominent note in the wrong position can ruin a phrase. He must, therefore, be prepared to reduce breath-pressure on a strong-toned note, or even re-finger it, so that it plays below its best tone and gives its neighbours the chance to be as or more important.

Variations in tone-quality: breath-pressure and vibrato

Flexibility of tone is a vital adjunct to variation of volume in achieving phrasing (that which transforms musical sound into music), and it is because so many players disregard both, that the recorder is sometimes thought to be capable of producing only an expressionless, emotionless monotone. The fact that the recorder has a small tonal range should be a challenge to its devotees to extract every drop of tonal variety their instruments will yield.

Assuming that there is no vibrato, and that breath-pressure is constant without any hint of changing or wavering, is it possible to change the tone of a recorder by changing the shape of the mouth? Both old and modern writers have suggested that tone can be improved by taking advantage of the potential resonance of the oral cavity. This can be tested by playing low and high notes, say F and D', in the latter case being sure that thumbing is unchanged, with two oral profiles. The first position is 'aa', as if you had a ball in your mouth; the tongue is at the bottom of the mouth, the lips are rounded but relaxed, the teeth well apart (i.e. the chin dropped, but not to so much as to break the lips' seal round the recorder mouthpiece), and the cheeks relaxed. This gives the maximum air passage between your throat and the recorder windway. The second position is 'ee', with the tongue placed high in the mouth near the upper part of the front teeth, the lips in a normal, not forward position, with the lip ends relaxed but very slightly smiling. This brings the jaw up, but make sure that it is not far enough up to cause the teeth to impinge upon the passage of air. It prevents any rounding of the cheeks – never play with the cheeks puffed out for this too can create unwanted eddies. Play into the corner of a room and listen extremely critically to the two sounds you make on each note, F and D'. Better still, play several notes to a discerning listener and ask him to write down, without knowing your order, any difference in quality he may notice in the note itself, ignoring its articulation phase. What matters is the sound *after* the note has been tongued. A further test is to start a note in one position and slowly move the tongue, chin, and lips to the other. Do you hear any change in tone-quality while playing the note? You might use an oscilloscope to see the harmonic pattern of the notes. I would expect the result of these experiments to be that you notice no difference in tone in the different oral positions.

There is no scientific reason why you should. Recorder sound is produced at the fipple edge, remote from the lips by the length of the windway and the distance between the windway opening and the edge. Resonance does not travel back into the mouth because, although there may be some residual vibration in the material of the mouthpiece, the soft tissue of the lips would effectively dampen this. The general breath-pressure used in recorder playing is low and the windway long enough to accomplish the full shaping of the air-column pushed up by the diaphragm. There can be no 'preforming' effect in the mouth (but see below).

Nevertheless, I advocate the use of 'aa' for full low notes, 'er' for middle notes, and 'ee' for high notes and for all notes played softly, for four reasons.

Firstly, we know that high notes call for more breath-pressure, although with a recorder as opposed to a flageolet or tabor-pipe, register-selection by higher initial breath-pressure is facilitated by the use of the thumb-hole as a speaker or octaving hole. This helps to prevent a high note from breaking downwards to the lower octave if breath-pressure is reduced after the note has been started. The 'ee' sound moulds a narrower passage of air through the mouth. We can therefore play a note at higher pressure but using less air (more volts, less amps). Using less air compensates for loudness; that is to say, we can then play high notes more softly, enabling us to overcome the natural disposition of the instrument to play high notes loudly and low notes softly. Conversely, the low notes, though lightly tongued, need filling out with air immediately after striking to give them a rich tonal quality at a higher volume so that they are on equal terms with the high notes. This is particularly important with a baroque recorder which is voiced in favour of the high notes.

Secondly, if the passage of air through the mouth is restricted by using 'ee', there is more resistance to the air passage, enabling the player to exercise greater control in maintaining an absolutely steady flow of air. The less air used, the softer the sound produced, and the more difficult it is to control a note without wavering. Test for yourself the gain in control by playing notes *ppp* with 'ee' rather than 'aa'.

Thirdly, consider the direction of the tongue stroke from 'aa' and from 'ee'. 'ee' enables the player to use a shorter, more lateral, and more direct tonguing. It is more precise and therefore ideal for articulating high notes. From 'aa' the tongue has to travel upwards and forwards to effect tonguing and is therefore flabbier. Thus it moves in a more rolling, grazing position, fractionally more backward on the teeth-ridge. This encourages soft, light tonguing which favours low notes.

The final reason, probably the most important of all, is psychological. Ganassi repeatedly enjoins his readers to imitate the human voice – 'the flow of breath is increased or lessened in imitation of the nature of words'. If you imagine you are singing into the recorder, or rather making it sing, you will be a better player. The oral configurations used by singers in obtaining their most beautiful tone should therefore be carried over into the way you put breath into your recorder, though do not attempt to imitate a modern singer's vibrato.

Reverting to the second reason for vocalization in recorder playing, one can audibly change tone (for the worse) by going beyond a normal 'ee' and deliberately shaping the mouth to impede the passage of air and offer high levels of resistance. This can be done by pressing the tongue forward up against the hard palate and teeth, or by closing the teeth, or by a mixture of the two. The change in the quality of tone can be detected by moving from an unimpeded 'ee' position towards a more restricting position. These positions cause sound-making eddies of their own: try the effect of using this type of breathing without the recorder. The eddies have a marked effect on the undertone of high notes such as D', but a lesser effect on low notes, with their weaker undertones. When playing softly, the eddies are less strong, and cause less tonal change. Moreover if the tongue is nearly at the 'r' tonguing position it has only to move fractionally to 'd' to cut off the note. In a full decrescendo (*a niente*) the closing of the jaws adds to the player's feeling that the sound is thinning out and disappearing. Despite the tonal implications, this is therefore a valuable technique. But it is imperative that, on resuming a normal level of volume, the tongue stroke should revert to its normal length, and the jaw be lowered to prevent the teeth impeding good tone-quality.

While it is true that each note gives its best sound at a certain optimum breath-pressure, some variations above and below that optimum will still make good sounds. Sometimes a fractional change in breath-pressure will alter the character of a note – a phenomenon more noticeable on the larger recorders. Beyond this, the higher breath-pressures produce an impure, reedier, and slightly edgy sound which has the brilliance to compete with other instruments and is therefore suited to sonata playing, while the lower breath-pressures give a purer, duller, and rounder note of a more ethereal quality ideal for recorder consort music. Both overblowing and underblowing must be recognized and avoided, as the one causes coarseness of tone and the other a reticent breathiness. An overblown note is not the loudest note because it is not all pure sound, and an underblown note, though undeniably the softest, threatens not to be a musical note at all.

It was to some extent the poverty or coarseness of notes blown at very high or very low breath-pressures, and the further tonal degradations caused by fingering to keep them in tune, that led to the recorder being superseded by the improved flute in the later eighteenth century. The control which a flautist exercises by his lips in sound production enables him to play notes of good quality both pp and ff: the latter in particular was an advantage as string instruments became louder and orchestras larger. But in the twentieth century the impure tones produced by a recorder at its extremes have entered the vocabulary of music. Some composers have even taken advantage of the sounds a recorder makes when the head-joint is removed from the body of the instrument and pitch control is achieved by hand positions over the bore joint – plaintive drooping glissandos, for example. Tone is affected by placing the hand over (but not touching) the window. 'Recorder harmonics' have already been referred to (p. 60). Many thumbed cross-fingerings combined with sharp tonguing will produce two notes at once (try 0 12– 45–7 for example), and changes in breath-pressure will then alter the balance of the two notes, making the upper or lower become more prominent. 'White noise' can be insinuated into a recorder's tone by allowing the lips to part slightly so that some air escapes. Other qualities of tone can be discovered by removing the mouthpiece just away from the lips and blowing into or across the windway. One can blow directly at the window, cross-flute style. Other sounds which some will regard as fascinating, others abhorrent, can be made by humming or singing into the recorder: the eddies created by the voice-production result in impoverished tone, but an eerie sort of two-part music can then be played by one ingenious player.

Rapid alternation of breath-pressures, that is, vibrato, is the recorder-player's chief means of obtaining tonal variety. He should be able to play without vibrato, and with fast, medium, or slow vibratos of differing amplitudes, and to move from one to the other at will, ranging from the calmest note to the most passionate. A phrase can be shaped by vibrato alone, the notes which it is desired to accentuate being accorded vibrato, while those in the trough of the phrase are given little or none. When variations in volume are combined with variations in tone, and the shape of the phrase becomes emphasized by the reedier and more prominent quality of the louder notes, the full expressive range of the recorder becomes apparent. As a general principle, vibrato is used on louder notes, which are usually the summit of phrases, and on long notes. A slight swelling of vibrato (i.e. making the beats of the vibrato slower and wider) adds interest to a long note and makes it more declamatory: this is particularly effective when a piece opens with a long note. Vibrato has the additional advantage of taking away some of the edginess of tone in a loud or high note. When the recorder is playing with louder instruments such as strings or piano, vibrato helps to give prominence to the recorder part. Slow music, which is the hardest to play, needs plenty of tonal variety, and it can be given full expression with carefully modulated vibrato to shape the phrasing. Vibrato should, however, be used with great reticence in baroque and renaissance music.

As we saw in Chapter II, vibrato is normally formed by the regular movement upon the air-stream of muscles in the back of the throat. Players of instruments which offer strong resistance to breath-pressure (e.g. oboists) can produce vibrato from the source of the air-stream, by moving the diaphragm muscles. The feeling can be simulated by placing the hand on the diaphragm and moving it rapidly towards and away from the body to produce a succession of 'h's on exhaling. Probably no recorder offers sufficient resistance for this form of vibrato to be effectively employed in recorder playing. But one can *imagine* a deep, slow vibrato (though it probably still all comes from the throat) in playing the slow movements of some modern sonatas, in recitative or rhapsodic style (*PB* Ex. 6), to give a cantabile effect.

While breath-controlled vibrato was used in the renaissance and baroque periods (there are references to it sounding like the bleating of a goat), finger-vibrato was much more commonly used to ornament longer notes. It seems that the finger immediately beneath the lowest closed hole was often used, but this was at the side of the hole impinging on the emerging air, rather than half-holing, which would usually have made a full rather than a close trill (vibrato on bottom F was achieved by shaking the whole instrument – try it). The little finger beating beside its half-hole will produce a more controlled vibrato of lower amplitude. Try it on C′, gradually moving the finger across to cover all its half-hole, then back again to reduce amplitude and revert to the unadorned note. Finger-vibrato flattens a note, so it is particularly effective to introduce it into a note which swells in volume, keeping the note in tune at the same time, possibly also with the help of some extra shading. A player with a renaissance recorder in his hands would probably do well to forget breath-vibrato and cultivate finger-vibrato. In baroque music finger-vibrato was still in the normal rule: Quantz warns against a trembling action in producing tone from the chest, and to control intonation in a crescendo advocates 'making a vibrato with the finger on the nearest open hole'. Vibrato produced by beating the finger on a more distant open hole, and upon the edges of open holes, is referred to by Hotteterre; he provides instructions for each note. Once you become accustomed to finger-vibrato you will find its tone-quality convincing, and fluctuations in speed and amplitude gratifyingly easy to control. But beside finger-vibrato as an ornament, it is probable that in singing, and therefore also in playing wind instruments in the baroque period, a light inconspicuous breath-vibrato from the throat was used in the expressive shaping of phrases.

The whole range of vibrato (including finger-vibrato where specifically requested) should be used fearlessly in the interpretation of twentieth-century recorder music.

Tongue-vibrato (y-y-y-y) is simple to manage, but there is no historical evidence of its use, although it could be used to simulate the early seventeenth-century vocal 'trillo' – a reverberation of sound with no fluctuation in pitch. It has great value as a form of tonguing (see above), but it is not readily acceptable to modern ears as a form of vibrato because it is too light to be convincing.

Variations in tone-quality: tonguing

Tonguing has relatively little physical effect on tone-quality, as it only affects the beginning of a note. It does have a psychological effect, as less precise tonguings prepare the mind for a softer tone to match their 'unctuousness'. A passage can be made to sound different by varying only the level of the tonguing, but the difference in sound is one of attack rather than of tone-quality. Sometimes, however, attack is so marked as to affect the overall quality of sound, e.g. 'chiff' in renaissance dances. Strong tonguing at low breath-pressure on staccato notes produces a damped and spongy texture effective in certain contexts. Bad tonguing, of course,can ruin the tonal quality of a piece: I have heard a whole passage of semiquavers played with a spit effect at the beginning of each note due to over-strong tonguing.

On the other hand, extremes of tonguing form part of the tonal vocabulary of avant-garde music. Some composers have asked for a staccato note with so powerful a tonguing that it is spat rather than merely articulated. Strange effects can be achieved by using tonguings such as 'th' and 'n', by placing the tongue directly on to the windway mouth, by tonguing against the lips, or by articulating a note by putting the mouthpiece to the front of the lips (rather than between the lips) and 'tonguing' – perhaps 'pouting' would be more descriptive – with the lips only. Even Ganassi thought up this latter idea.

Variations in tone-quality: fingering

Fingering plays a more important part in achieving variety of tone in playing recorders than is the case with other wind instruments, and the references to tone in Chapters V and VI illustrate this. Differences in tone-quality are caused by differences in the intensity of the various overtones and subsidiary notes that go into the making up of any note. The breathy undertone audible below thumbed notes on the recorder can coarsen tone-quality if the position of the thumb makes it discordant with the main note, and, with two or more instruments playing discordant difference tones, can produce a coarse and unpleasant sound. The cross-fingerings so often used as alternatives have a pattern of harmonics different from plain-fingered notes, and produce a less dominating, more constricted quality of sound extremely useful in soft, contemplative music when their comparative thinness of tone gives a pleasing effect of distance. In consort music, these cross-fingerings are valuable in background parts when another player has the lead; they become essential if the notes in these parts are naturally rather prominent ones such as A on the descant, which in these circumstances would be fingered 0 1–3 45$\cancel{6}$–. Other alternatives have a stronger and coarser tone than the normal fingering (e.g. treble alternative G′ – 123 4567). These tonal variations of alternatives should be exploited in giving expression to music, as well as in avoiding obstacles to good expression sometimes inherent in the use of ordinary fingerings, for example the unwanted prominence of an A♭′ in a C minor scale passage, or of an E♯ leading to an F♯′, (better fingerings: Ø 123 45$\cancel{6}$–, and – 123 –––– respectively). The demands of the music may require a thin and distant D instead of the firm and beefy tone of the normal D: the alternative provides it.

Even admitted defects of a recorder can be pressed into service when the context justifies it. The clicks made in changing registers sound very effective in a piece of sopranino bird-music (e.g. the A′ to G′ slurs in *Le Rossignol en Amour*), and the sharp trills a recorder sometimes tends to perpetrate (e.g. B♭ to A) have an arresting tone-quality most suitable in a virtuoso solo piece. Devices intended primarily to improve intonation and volume control also affect tone-quality. The 'wheelbarrow' sound-projecting gadget and Plasticine 'wings' make tone more penetrating and rasping, while the opening of tuning holes in the head-piece produces a very slight breathiness in tone-quality. Shading the foot of the instrument constricts tone and gives it acidity.

As recorders are not (at least in their smaller sizes), particularly expensive, players can widen the tone-colours at their disposal by the simple expedient of possessing two or more descant and treble instruments. The clarinet-player, after all, possesses a B♭ and an A instrument for ease in managing various keys: why should the recorder player not take advantage of the great differences in the tone-qualities of instruments built on different principles, and possess a solo and a consort instrument? The former would be a baroque-type instrument of conical bore and voiced to a high breath-pressure, so having a penetrating reedy sound capable of joining forces with keyboard and strings, while the latter would be a more cylindrical bore renaissance instrument which, with a lower breath-pressure, produces a round and ethereal tone suited to the calm of consort music.

Given an instrument able to create beautiful sound, and a player with the technical knowledge to get it, there is only one more condition necessary for its production. And that is the right attitude of mind. No player will produce good tone unless he knows what he wants and believes in the goodness of his instrument and the worthwhileness of the music he is playing. Conviction will carry a recorder-playing far. If he truly believes that the instrument he plays is 'une flûte douce', the chances are that it will be so.

IX
ORNAMENTATION

While a recorder-player whose enjoymentof his instrument lies
in consort and recorder ensemble music can get by with no
knowledge of ornamentation, one who proceeds beyond this
point to play baroque sonatas must begin to study this difficult
and controversial subject. It is difficult because musical style
has changed continuously down the centuries, sometimes
rapidly, sometimes less so. Although music is a common
language and composers moved freely from one European
country to another, each country has its own national
characteristics and musical style, so a mode of interpretation
that may be correct in Italian music in 1700 may not be so in
French music of the same date. Two styles may overlap at one
time, however, or even within the works of one composer, e.g.
Purcell wrote Fantasias for strings in a retrospective English
style, while writing more Italianate vocal works, and Bach
wrote both French and English suites and the Italian Concerto
for harpsichord. The subject is further compounded because
writers on ornamentation do not agree among themselves, even
in the same country or the same period. Therefore if one tries to
derive general principles, one has to accept that they may be
incorrect in a substantial minority of their applications. But one
has to start somewhere.

There are excellent books on ornamentation, many of them
being the result of profound scholarship during the last two
decades: a critical selection is given in Appendix 2. Some are
supported by extended examples of a composer's own ornamentation of his compositions (we thereby know a great deal
about Bach's and Telemann's ornamentation). One of the
soprano songs written by Antonio Archelei for the Florentine
marriage festivities of 1589 was embellished by his wife Vittoria,
one of the foremost virtuosi of the time, and the publisher,
Malvezzi, printed her ornamentation. Ganassi's *Fontegara*
(1535) is largely a treatise on free ornamentation for the
recorder. At least from the sixteenth to the eighteenth centuries

there is no shortage of source material. In the nineteenth century, composers expected performers to play what they had written, liberty only being granted in cadenzas. Today some composers, for example, of aleatoric music, have reverted to allowing performers liberties to such an extent that it is difficult to know within what conventions, if any, the player is bound.

Gramophone records illustrating ornamentation are available (though unfortunately, one of the best sets, Denis Stevens' *The Art of Ornamentation and Embellishment in the Renaissance and Baroque* (VSL 11044–5) has been deleted), and many soloists and early music groups cultivate authenticity in performance practice. One of the best ways of learning about interpretation and ornamentation is by thorough and frequent listening to such performances, preferably with score in hand.

This chapter must therefore deal with ornamentation in a rather perfunctory and highly generalized way: you must go to specialist books on the subject to get the information you need to give adequate performances of baroque music. It will, however, give a little more attention to the technique of playing ornaments on the recorder.

Renaissance ornamentation

The following general statements may be taken to be true for ornamenting music of the Late Renaissance (approximately the latter part of the sixteenth century).

1 Ornamentation is a more or less continuous process of embellishing, usually in semiquavers, music written in longer notes.

2 The process is based on formulas for dividing into semiquavers different intervals in the original melody. 'Memorise enough patterns in nine days, one day each for ascending and descending seconds, thirds, fourths, fifths, and one for unisons' (Conforto, 1593). These formulas were a legacy from the fifteenth century, perhaps earlier, of the processes of free improvization of complete parts to a piece (e.g. a *basse danse*).

3 The movement of the ornamentation is steady and measured (this is an uncertain generalization as Ganassi divides semibreves into 7s and 5s; but other writers are less rhapsodic).

The semiquavers can themselves, however, be divided into shorter notes, or mixed with shorter notes (more true later in the sixteenth century).

4 The ornamentation must embellish the original melody, so that the main melody notes (i.e. those at chord changes) must be preserved. Passing notes in the original melody, however, may be ignored.

5 The ornamentation should itself be melodic in character, creating something new which, however, relates to and comments upon the original melody (the Spanish word for ornamentation is 'glosas' = gloss or commentary).

6 This new melody will tend to weave round, above, and beneath, the original melody note. It can, however, take the melody note up to an octave higher (or lower).

7 In ensemble music parts do not ornament simultaneously (they do occasionally overlap in thirds in the process of making a continuous semiquaver web in the texture).

8 Each note must be articulated independently, even in faster-moving sections (see *3* above).

9 Only three specific ornaments ('graces') were regularly used in renaissance music, though some special ornaments are associated with e.g. the voice or the lute:
mordent rapid movement to the note (a tone or semitone) beneath, or sometimes above, the note being played. Mordents may be shown by the sloping sign used in renaissance keyboard music (ornament signs are not otherwise much used in renaissance music).
tremolo rapid trembling (= a very fast trill – was it slurred?) usually taking up half the main note, more often the first than the second half. This very fast trill starts on the main note and can alternate with a semitone, tone, or even a third above, or somewhere between. Ganassi refers to a narrow tremolo as tender (*soave*), a wide one as lively (*vivace*). In a passage of descending equal notes, the first and every other note might be accorded a tremolo. A tremolo on a semiquaver is equivalent to an inverted mordent.
groppo a prepared trill, being groups of semiquavers weaving in the manner described in *6* above, but leading on to the final note

of a cadence by speeding up to eight demisemiquavers. The rhythmic emphasis of the eight demisemiquavers, which were tongued if possible, is on the note above the leading note, i.e. the upper auxiliary, as in most baroque trills. The last two notes were turned. The trill section of the *groppo* could be approached from below, i.e. (b) rather than (a)

10 Renaissance music can be played without ornamentation.

Towards the end of the sixteenth century, dotted notes began to feature more in patterns of ornamentation, and in addition, ornamentation became more rapid and showy – some said excessively so – by applying the principle referred to in *3* above. It thus became necessary for the sake of musical sense for some notes in the original melody to be left unornamented, so breaking up the stream of ornamentation and concentrating it upon elements of the melody. These shorter sections of faster ornamentation developed during the early seventeenth century into the more schematic ornaments of the baroque period. To quote Donington (see Appendix 2) 'An ornament is a short melodic formula which has formed in the tradition of free ornamentation as a crystal forms in a saturated solution.'

Baroque ornamentation

Many of the principles of renaissance ornamentation, particularly in free ornamentation, apply also to baroque ornamentation. But two major changes occurred. The first was that the process of 'crystallization' developed to such an extent that much ornamentation in the baroque period is in the form of specific ornaments, known under a confusing variety of names in different languages. French baroque music depends for its effects on the tasteful use of specific ornaments, and many French composers preface their music with a table showing how to execute the ornaments represented by various signs.

The second change was the development of figured bass and with it the appearance of the appoggiatura, a discord against the concord represented by the figures of the thorough bass. Its dramatic use, especially in cadences, is a feature of baroque opera and instrumental music, and it replaced the mainly concordant semiquaver preparation for the trill in the renaissance *groppo*.

We can now make some generalizations (i.e. more often true than false) about baroque ornamentation:

1 Baroque music cannot be properly played without ornamentation.

2 Some ornamentation, such as trills at final cadences of slower music, is obligatory, whether marked or not.

3 Further ornamentation is expected on repeats.

4 Ornamentation is an essential element in shaping phrases, to give emphasis to important notes; it enhances, not overlays, the melody.

5 One of the most important ornaments is the appoggiatura. This starts on the beat, and normally occurs from above the written note, though it can be played from below (see also *10* below).

6 In common or duple time the appoggiatura at cadences takes half the value of the main note that follows it: in triple or dotted measures, it takes two-thirds of the value.

7 Trills are played slurred and fast, but at a speed to match the mood of the piece, i.e. slower trills in a *Largo*. All trills in the same piece, or section, should approximate to the same speed. Other than short trills, however, trills are not precisely measured: they may in some contexts speed up slightly towards their conclusion, if they last long enough.

8 Trills start on the beat, and with the upper auxiliary, which remains dominant during all or most of the trill.

9 Trills resolving upwards arc turned. Passing trills (i.e. trills not on cadences) are turned if there is time (i.e. usually). It is especially important to turn them if they fall on an unaccented note (e.g. the second or fourth note of a 4/4 bar), even if, because of the speed of the piece, this means playing only four notes. Whatever is written in the music, the turn occurs at the same speed as the end of the trill (though it can get caught up in a *rallentando*). Cadential trills resolving downwards are not usually turned:

10 While appogiaturas at cadences, and mordents or half-trills, start on, not before, the beat, much baroque ornamentation consists of lightly played notes between the beats of a melody. The slide or *coulé* (or passing appoggiatura) connecting descending thirds, takes place before (and is slurred on to) the note on the beat. Ornamentation was played with great individual freedom.

11 Ornamentation which moves through a succession of notes should take account of the prevailing harmony; e.g. a flourish should be in scale-notes of the key of the piece and should change direction only on a note belonging to the chord being played. Chromatic runs only became common in the classical period of the later eighteenth century (e.g. in Mozart piano concertos), and ornamentation in triplets belongs to later rather than earlier baroque practice.

The purpose of ornamentation

The prime purpose of all ornamentation is to make beautiful music yet more beautiful by bringing out its shapeliness or emphasizing its liveliness. Ornamentation should arise out of an appreciation of the qualities of the music being played. Ornamentation has a secondary effect of making the music even more pleasing, or more exciting to an audience.

The display aspect of ornamentation *is* secondary, and this should affect your approach to playing ornaments. Never push them. Remember ornaments have an element of discord. This is usually sufficient in itself to attract attention. In particular, do not deliberately accent appoggiaturas, which are always discords. But eighteenth-century writers are inconsistent in this respect – some say place a small accent on the appoggiatura to help your audience to feel that it is a discord; others say place a slight emphasis upon the following note to announce your arrival on the concord. Compromise is generally the right answer – let the appoggiaturas just speak for themselves, taking whatever weight falls upon them in the natural musical rhythm. An appoggiatura should be stressed, therefore, when it occurs on a note that would normally be stressed if it were not an appoggiatura, for instance, a note in a hemiola (i.e. the treatment of two bars of triple time as if they were three bars of duple time). For an example, see the end of the second movement of Handel's C major Recorder Sonata.

Some ornaments, such as the mordent, have a bite of their own. Again there is usually no need to emphasize it. If you find you are emphasizing ornaments in your playing without good musical reason (your teacher will tell you if you are) deliberately unstress them.

Ornamentation may take place between notes (or even replace notes that are not essential to the harmonic or melodic framework). It should then be played lightly, with delicacy and freedom. Or it may belong to a particular note, often with an ornamentation sign above it. It should then be played with rather more precision and deliberation.

It is not easy to lay down general rules as to when and when not to ornament, as so much depends on the style of the piece. Early eighteenth-century French music, which is often not very melodious, depends for its effect on its being performed with 'taste and propriety', in other words, with appropriate ornamentation, and therefore demands more frequent and deliberate ornamenting than a piece in the Italian style in which ornaments are less frequent and more subservient to melody. On the other hand Italian-style Adagios are often an invitation for considerable free ornamentation, as shown by Corelli's ornamentation of his own slow movements, and, much later, by Quantz's long chapter 'Of the Manner of Playing the Adagio'. The effect of ornamenting a given note is to draw attention to it: it has the same result as a dynamic accent or as vibrato (which is a sort of ornament). It should therefore be related to phrasing, decoration being according to notes in a phrase which need bringing out. A secondary effect of an ornament, particularly one ending with a turn, is to give music a forward impetus; this is partly because part of every ornament is a discord and the ear anticipates its resolution. The leading note of a cadence is generally improved by decoration, and except in fast pieces its decoration at the final cadence of a whole section of music is obligatory. 'Embellishments', to quote C. P. E. Bach, 'make music pleasing and awaken close attention'; the variety they lend to a repeated section of music, played first only with cadential trills and such shakes and mordents as are essential to the phrasing, is particularly delightful. Slides and flourishes are particularly effective in Sicilianas and other movements in triple or dotted time. Rapid trills and flourishes also have the effect of making music more exciting. This quality is both an advantage and a danger. It is good to be able to compel the attention of one's audience, but sheer pyrotechnics, which often sound much harder to execute than they are, can make nonsense of music. The criterion to adopt is, 'can each ornament be justified musically?'.

References are made in Appendix 2 to books which give examples of ornamentation, often from original sources. As a starting-point, however, study the second half (pp. 37–58) of Freda Dinn's *Early Music for Recorders* which not only gives fully written-out examples of ornamenting whole movements, but in the text explains why particular ornaments have been used in particular contexts. She takes the reader through Handel's G minor Recorder Sonata, the Sonata da Camera by Thornowitz, and a French suite by De Caix d'Hervelois. Next extend your knowledge of French ornaments by reading and playing Leonard Lefkovitch's edition of Hotteterre duets with the composer's own explanation of the ornaments. Note that Hotteterre treats vibrato – finger-vibrato – as an ornament ('flattement') and says it should be used 'on nearly all long notes, slower or quicker according to the tempo and nature of the piece'. His *Principes* show how to finger it on each note of the recorder; these fingerings indicate the use in his time of quite a wide amplitude of vibrato. At this point you will be ready to broaden your knowledge of ornamentation by reading, and playing examples, from the more thorough-going books on ornamentation listed in Appendix 2.

For making a start on renaissance ornamentation, I advocate playing the ornamented descant part in Bernard Thomas's edition of fifteen dances by Demantius. Having got the 'feel' of this kind of ornamentation, study it more thoroughly in Howard Mayer Brown's excellent book on the subject.

When you play music with much ornamentation (e.g. renaissance divisions) try and keep the original melody in the back of your mind. However delighted your audience may be with your decoration, it is the original melody they should go away humming.

Execution of ornaments

The best way to begin learning how to play ornaments is with the turned trill. Practise slowly the second trill on p. 116, first as it is written, on the note D (i.e. E–D–E–D–E–D–C–D–E). It should be enunciated 'Dhee-er-ee-er-Ee-er-aa-er Dhee', perhaps using 'y' tonguing in accurate synchronization with the finger movements as an extra method of control. This is the slurred baroque passing trill. Now play it with light tonguing for renaissance use – 'Dhee-rer-lee-rer-Dhee-rer-laa-rer Dhee'. This is the first trill on p. 114; but practise both examples – the second starts 'Dhaa-rer-lee-rer . . .' etc. Now repeat the exercise on F', i.e. starting with G', the note above, and using alternative fingering for the E turn (*PB* Ex. 57). Next try it on C (starting on D), and on upper B'. Now trill on E with the normal alternative for E and two fingers below that (and a slight drop in breath-pressure) for an alternative D turn. The A' trill needs an alternative G' turn and the thumbed, one-and-a-half below, G' should be suitable. The other main G' alternative (all on except thumb) is used for the G' trill, with A', its upper auxiliary, also being an alternative (– 123 4–67 – thumbed to start the trill), but for the note preceding the F' turn, the sixth in the sequence, an ordinary G' must be used. This needs careful practising, but it must be mastered as the G' trill is both frequent and hard to control, firstly because of the waywardness of the third finger and secondly because the tonal strength of the alternative G' tends to upset the pattern of the trill. A whole C major scale of trills has now been worked out, with articulation in both the baroque and renaissance fashions. These trills should be practised at ever-increasing speeds, though always kept neat, even, and strictly *a tempo* (practise with a metronome if possible). The same exercise should now be carried out with fast six-note turned trills, and with longer trills of up to sixteen notes. Speed is essential.

When the C major trill scale is mastered, start at the beginning again with the Ab major scale. Certain trills in this scale cause new difficulties. The Bb to Ab trill itself needs an alternative Bb (0 123 4–67) to start it but as the trill with 5 and 7 is awkward, 7 may be abandoned causing the trill to be slightly sharp. This may be effective as the sharp trill has an acidity of tone that helps it to stand out, for like most low notes of the recorder it lacks brightness of tone, and the possibility of increasing volume is limited because of the low breaking-point of the forked fingering. If the context requires the trill to be accurate in intonation, the 5th finger can still produce the trill alone by staying low over its hole while it trills, i.e. shading the Bb of the trill. Similar conditions affect the difficult Bb to A trill (see p. 71 and *PB* Ex. 59). Fingering problems arise literally at every turn, but experiment and reference to Chapter V should solve them. The general principles are:

(i) Whenever possible find a way of trilling with one finger; it is better to trill slightly out of tune with one finger than to fluff a trill with two.

(ii) If this is not possible, keep the trill to the fingers of one hand.

(iii) Use an alternative for a turn even if it is a poor quality note and requires a drop in breath-pressure.

When starting to learn to trill, trills should be executed with a hammering action and, unless there is need to shade the trilling hole, the trilling finger(s) should be lifted high between each blow: at this stage you may find that with light fingering, trills tend to speed up and run out of control. Although the trilling finger is somewhat tensed in its hammering movements, the other fingers should remain as relaxed as possible, all energy being concentrated, as it were, into the trilling finger. At a later stage, when you are confident that you can manage all trills with precision, you should make your fingering movements in trills and other ornamentation as light and economic as in your recorder playing generally. For trills on half-holes (e.g. A to G♯) swivel the wrist back so that the trilling finger remains in a comfortable position rather than bent up.

Once the turned trill is thoroughly mastered, other schematic ornamentation is simple if it is regarded as a part of, or as an extension of, the turned trill. The cadential trill (as on p. 116) requires closest concentration: the appoggiatura should only be on an alternative fingering if the tone-quality of the alternative is good, so a rapid change in fingering is often needed for the trill itself. Similarly, the end note of the trill (i.e. the written note), should be good and not a weak alternative, for it is dwelt upon long enough for poor tone to be apparent. The final short note should always be very short and should be played with the 'g' of double-tonguing: the pause for articulation before it should be longer or shorter according to the jerkiness or smoothness of the music. A common fault is to start the cadential trill a demisemiquaver too early. This can be eradicated by thinking 'Dheee . . .' for the appoggiatura, then '. . . ee – er – ee – er . . .', etc. for the trill, keeping the weight on the upper auxiliary. The appoggiaturas of cadential trills are not held on so long where cadences occur at the end of phrases in the course of a piece, especially if the music is fast. The length of an appoggiatura depends very much on its context. It can be short; in fact the first note of a passing trill may be regarded as an appoggiatura – the shortest one possible.

A mordent is taken 'Dhee – er – eee . . .', or 'Dher – aa – eer' for low notes; it is rapidly tongued 'Dhee – rer – leee . . .' in renaissance usage. A turn is equivalent to the last four notes of the turned trill, but with the last note held on. Mordents, shakes, and turns are usually played diatonically, though semitone chromatic mordents are also common.

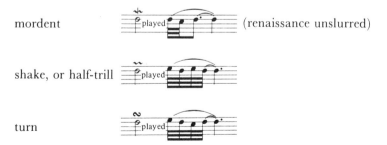

mordent   (renaissance unslurred)

shake, or half-trill

turn

In modern music, trills begin on the lower note except in pieces written in the style or form of old music, e.g. Herbert Murrill's Sonata.

The technical advice I have given in this chapter is no more than a beginner's guide to ornamentation. It is designed to help you through the first essential stage of acquiring control in ornamentation, with clean and accurate fingering and tonguing. If you do not feel a sense of control over your ornamentation, if you do not know where you are, the effect will be sloppy and ugly. You must first master the basic grammar, and then gradually gain experience of using the elementary patterns suggested in this chapter in your sonata playing (*PB* Ex. 62). But to apply only these basic formulas would be far too rigid an approach to the supple and ingratiating art of ornamentation. As your control and confidence develops, and as by listening critically to good examples and by reading the books listed in Appendix 2 you gain a greater historical sense, you should begin to be more flexible, adventurous, and imaginative, though always within the bounds of the melodic and harmonic structure of the music, and above all in a manner that interprets the basic mood or 'affekt' of the piece. With much practice and experience, your ornamentation should become relaxed, *soigné*, and as natural as if you were inventing it as you went along. Only then will it sound authentic.

X

PRACTICE

Amateurs generally have neither the time nor the incentive to spend long hours in regular practice, but on the other hand no self-respecting amateur could be satisfied with not being able to play music as he wants it to sound, and the only way of acquiring the technique upon which expression depends is by practice. Many players spend time simply playing through one piece of music after another: enjoyable though this may be it is less profitable in improving one's playing and understanding of music than systematic practice and concentrated work on a single composition, a process which eventually gives deeper enjoyment than aimless sight-reading.

An amateur recorder-player's practice should arise mainly out of working at a selected piece of music and the desire to interpret it sensitively. Technical difficulties present themselves in almost any piece, for even apparently easy music becomes hard if perfection is aimed at. Hypercritical listening to one's own playing of a simple tune brings salutary realization of deficiencies in intonation, tone-quality, volume, and variation and control of vibrato, while in more complicated music, such considerations as unevenness of fingering in obstinate semi-quaver passages, 'clicks' on slurs, and untidy ornamentation are added to these aspects of technique. So long as nothing short of the highest standards is accepted, the preparation of a piece of music for performance will reveal technical weaknesses, and it is upon these that practice should be based. The difficult passages in a piece should be ringed in pencil for special attention: play the passage slowly with meticulous accuracy, then gradually work up speed. When mastered, you can rub out your pencil marking. It is a less rigorous code of practice than would be followed by the professional, but as enjoyment is the be-all and end-all of amateur playing the improved playing of a chosen piece of music is more satisfying than the drudgery of fundamental practice with its less immediate results. The clue to success in such a practice method is to choose for performance only pieces which are within one's capabilities. But if you feel technical exercises will help you, and you enjoy the challenge of, for example, playing rapid F♯ major scales, by all means mix them with your work on 'real music'.

Breathing and tonguing

A well-known oboe player due to play in the St. Matthew Passion once asked to be provided with a studio for two hours' practice: he spent the whole of this time playing one note, over and over again, loud, soft, distant, commanding, with all grades of vibrato from the plain statement of fact to the most passionate and dramatic, notes which started plain and soft and worked themselves up to a frenzy of volume combined with wide, slow vibrato which narrowed and speeded up until the note died away in quiet calm. The chapters on breathing and volume (Chapters II and VIII) show how practice of this sort can be planned. Long-note practice (*PB* Exx. 1 and 2) gives the player the opportunity of knowing the character of each separate note on his instrument – its tone-quality and how it varies with differing breath-pressures, the amount and kind of shading it needs at different volumes, and, in relation to other notes, its intonation and how best it can be controlled (Chapters IV, VII and VIII).

The tonguing practice mentioned in Chapter III of playing Bb with the fingering 0 –23 4567 is aimed at encouraging light tonguing and it should be used regularly as a reminder of what light tonguing is. Other exercises are suggested in that chapter (also *PB* Exx. 9–17). Technical difficulties involving tonguing may arise from music with passage-work, with high notes, or, worst of all, with passage work on high notes (as in the Second Brandenburg Concerto – *PB* Ex. 46), where tonguing, thumbing, and breath-pressure must be practised until all three are right on every note (Chapters III and VI).

Fingering

Scale passages are so frequent in recorder music that the practising of scales out of context, although desirable, is not a necessity. Certain common sequences of notes, particularly those involving Eb or Bb, tend to unevenness because of the disparity of finger movement; a three-finger move, especially if it involves the fingers of two hands or the sluggish third finger of the left hand (the finger that Schumann tried so disastrously to strengthen), tends to take longer than a one-finger move, and a one-finger move on a first finger can easily be skipped in a fast scale passage. 'Five-finger exercises' (from tonic to dominant and back) in keys such as F major, C minor, D major, A major, and F♯ major, help to even out awkward sequences: they should be played separated, portamento, and slurred. Purely physical exercises, such as the independent moving up and down of those culprits the third fingers, or the sudden clenching and unclenching of the hand, spreading the fingers as wide as possible when unclenching and pressing them tight together when clenching, balance the strengths of the fingers. Softening the web between the fingers with olive oil and massaging the back of the hand between the knuckles also helps: if the fingers get tired with recorder playing, this rubbing, with the hand held absolutely limp, relieves tension. A player who habitually gets finger fatigue is almost certainly not relaxing his fingers and he should deliberately untense the finger muscles so that he becomes aware of the dead weight of his fingertips. But finger fatigue may simply be caused by practising for too long at a time – Quantz warned against too much practising.

When speed has been gained by hammer-blow fingering in the early stages, fingering should gradually become lighter and looser, though no less rapid in movement; eventually the fingers should rest so gently on their holes that the vibration of the air in the instrument is felt on the pads of the fingers. In the hammer-blow stage of learning to finger, the fingers must perforce be held high, but a practised player who holds unused fingers high is wasting effort in making his fingers travel farther than they need. If the fingers have farther to move, fingering is bound to be slower, so the ideal to aim at is to hold unused fingers just above the point where they begin to cause shading – if they are held too low they may unwittingly flatten a note. As a general rule the unused fingers should lie in a plateau not more than an inch from their holes. The weight of the instrument should be taken entirely by the right thumb against its thumb-rest. The feeling in the fingers of a practised player should generally be of buoyant lightness, rapidity, and independence of movement, of fingers dancing over the instrument yet scarcely touching it.

Passages of semiquavers should trip along gaily, the fingers lilting with the music.

Practice is the opportunity for experiment, and a player faced with a difficult passage to master should see whether it can be made easier by the use of alternative fingerings. Where an alternative fingering is available, practise it until you are familiar enough with it to make a fair judgement as to whether it is better than the normal fingering: when the difficulty is not significantly relieved by the use of the alternative, choose the normal fingering. The use of alternative fingerings, especially in fast passage-work, trills, and turns, and music where a wide dynamic range is called for, can save a lot of unnecessary practice. Even apparently useless alternatives may be pressed into service in certain contexts: for example, in extremely rapid slurs, top or bottom Bb may be played unforked, or in trills from C to Bb the first finger of the right hand can do the trilling while the fork stays down. The chromatic run in Lennox Berkeley's Sonatina (*PB* Ex. 17) can be approached by using 0 –23 45–– for D, 0 –23 4–––– for Eb, and 0 –23 –––– for E, instead of spending effort on attempting to contrive a very rapid chromatic run in ordinary fingerings. The principle to go on is to use the fewest possible fingerings for each note change: economy is the goal, though false economy must be recognized and avoided.

Suggestions for the practice of trills and other ornaments are given in Chapter IX (and *PB* Exx. 56–62).

Thumbing

Sudden jumps or slurs over wide intervals are difficult on most instruments, and the recorder is no exception. Success depends on the rapidity of thumb movement and control of breath-pressure (see Chapters II and VI), and the secret is in the position of the thumb when it is closing its hole. It should then be so near to the octaving position that a fractional bending of the thumb-joint will bring it there: in fact the thumb-nail should touch or nearly touch the instrument even when the thumb-hole is closed. The thumb movement can thus be made quickly, a slightly increased stream of air put into the recorder, and a slur of an octave or more is accomplished with ease (*PB* Ex. 35). The thumb must move quickly, but at the same time be completely relaxed. If it becomes tense with lack of confidence it cannot do its job so well.

Scales, and all fingering and thumbing practice, may profitably (especially to a player who does not frequently play with a group) be carried out to the accompaniment of a metronome, however exasperating this practice is. But it would be masochistic to do all your practice with the aid of a metronome.

Sight-reading

Apart from the technique of playing the instrument, the recorder-player has to master the technique of reading music at different pitches and in various clefs, otherwise he is not fully equipped for playing consort music (*PB* Exx. 63–5). A treble player should be able to read at pitch (involving familiarity with leger lines), and from music printed an octave lower – playing descant music on the treble is a good exercise in 'reading up' and in negotiating high notes: in consort music published with two viola parts, the treble player may find himself having to read the alto clef an octave up, his bottom F being on the lowest line of the stave. A tenor player may be asked to use a viola part, and he should also be able to read from the bass clef in close score. The bass player must be able to play from his usual bass clef, and from treble clef both at pitch (i.e. with leger lines below the stave) and an octave down when music arranged for that shrieking combination two descants and treble is made to sound more beautiful on two tenors and a bass: he must also be adept at choosing the right moment to move an octave up when notes lower than F are looming in his score. Recorder-players may at least be thankful that the French violin clef is no longer in use, but its acquisition (playing treble recorder as if reading bass clef – for G is on the bottom line) opens up the possibility of reading early eighteenth-century music from facsimile, thereby getting as close as possible to the composer's intentions. Some musicians evolve fascinating short cuts to cope with different clefs, but for the recorder-player the best way of approaching a strange clef is to locate two or three notes to act as 'anchors' and then to read by intervals as a singer does. Thus a tenor player reading the alto clef locates G, C′, and G′: the piece starts on D′, goes up by a third, then down a fourth, when the player checks he is playing C' (middle line). Common accidentals such as B♭ and F♯ also assist the player in finding his way around. An excellent exercise is to sing the piece to the names of the notes. This 'interval' method is better than any 'mechanical' process of reading, because the player must hear the interval in his mind, and this 'pre-hearing' of sound – knowing in advance the note, or sequence of notes, which must be produced – is the secret of good sight-reading. In sight-reading practice, deliberately keep the eyes ahead of the notes that are being played, and if they slip back force them forward again. Get a friend unexpectedly to snatch the music away from you while you are playing and see how far you can go on without it. The ability to read well ahead makes awkward turn overs more manageable. It also gives you just time enough to move over to a common alternative to avoid a 'clickish' slur (*PB* Exx. 38–9). Naturally, though, you would not use the more esoteric alternative fingerings in sight-reading – you must make do with normal fingerings for most of the time.

Take every opportunity of combining practice with pleasure. Use good music as your basis of practice: if you exhaust the technical difficulties in Handel and Telemann (and you will be a good player to do that) appropriate Bach's flute sonatas as recorder music, for the remoter keys – one movement is in C♯ minor – offer all sorts of new difficulties. Country dance tunes, and especially those from Ireland and Scotland, provide excellent material for finger exercises, and compel one to maintain a firm rhythm at the same time. To test the efficacy of your practice play the recorder for some country dance society – extra musicians are generally welcomed. Above all, be purposeful. The singling out of imperfections and their systematic elimination is the only way to progress. Your private practice should be such that if at any time someone interrupted you and asked to what purpose you were practising, you could give a cogent and unashamed answer.

XI

PERFORMANCE

Performance is not necessarily public, nor indeed need there be an audience at all. It is an attitude of mind, the putting over of the finished product. There are three stages in playing music – reading, rehearsal, and performance. Reading is the process of familiarization when an attempt is made to hear the music as a whole and to find what it is about; rehearsal is the section by section analysis of the music when decisions are reached on details of interpretation and how this interpretation may be expressed in terms of the technical potentialities of the instruments being used; performance is the final result, the exhibiting of music to a real or imagined audience. In performing music, the player is more emotionally alert to the music, for by then the reason for every turn of phrase and its relation to the meaning of the piece as a whole will have been worked out: in performance the players' understanding becomes the audience's experience.

In choosing music for performance it is important, therefore, to select a piece that is within one's understanding, and not of such difficulty that that understanding cannot truly be expressed. In fact, as far as technical difficulty is concerned, the music chosen should be such that the player regards it as 'easy' (a standard of difficulty varying with his technical proficiency); then his mind will not be distracted by questions of technique when he is actually performing the piece. The music chosen should be a piece the player likes well enough for him to want other people to hear and like it. It should also be one which, if there is a real audience, is likely to appeal to the tastes of that audience – their best tastes.

Preparation of music for performance

There is a definite order in which the preparation of music for performance should be carried out.

Style. First read the music (not necessarily playing it) to discover what it is about. Unless it is 'programme' music its real meaning will be in musical terms, but some attempt at extra-musical categorizing should be attempted. Most music, for example, falls into one of the three categories of song, dance, or narration. If the music is songlike, that is to say amenable to a verse pattern of words and containing relatively few wide intervals, a clue to its interpretation is already given – it should imitate the flexible movement of poetry with groups of notes articulating the syllables of individual words. If the music is intended to accompany dance, certain notes should be played much shorter than their written value to give 'lift', and the rhythm should be deliberate and forward-moving so that it carries the dancers with it: in this connection it is important to know something about the steps of old dances (see Mabel Dolmetsch's two books *The Dances of England and France* and *The Dances of Spain and Italy*). If the music is in an extended narrative form, it must be thought of as a complex of words, sentences, and paragraphs, of statements developed and carried to a conclusion. Music may easily have elements of each of these categories, but, more often than not, one predominates.

A second method of approach to music is to discover its prevailing mood; one should be receptive to any emotion it might express, or to a 'programme' or series of pictures or events it might suggest, for imagination engenders feeling, and to feel something about a piece of music leads to good phrasing. Such an approach accords with the eighteenth-century theory of 'affekts' in which music is seen as conveying an emotion or abstract quality (e.g. 'boldness') to the audience. External indications such as the composer's title of a piece, or its context in, say, a cantata or an opera, provide a valuable guide to the mood of a piece of music.

Music can further by categorized by its period and its style. If a piece of eighteenth-century music can be recognized as, say, an overture in the French style, the player who (as every recorder-player should) has listened to and read about the music of that time, knows at once how to play it: dotted notes are held on, semiquavers shortened and double-tongued, the movement made slow and lurching, and the sonority noble and pageant-like. Is the whole piece in the French style? If so, the manner of ornamentation will differ from the Italian style, and the whole question of inequality will arise. Books such as

Donington's (see Appendix 2) are essential to elucidate interpretational and stylistic matters. To know about music and, before playing a note, to think systematically about each piece that is to be performed, is the secret of playing it well.

The next stage in one's train of thought is to consider the *speed* of the piece. If the music is in dance form the steps of the dance may decide the speed (although some dances such as the Sarabande varied in speed between different periods): if the music is songlike it cannot be too fast for the proper articulation and expression of words. The time signature, taken in conjunction with the sometimes misleading Italian speed indication (Allegro, Largo, etc.) and with the nature and frequency of the shortest-value note in the piece, supply the remaining objective guidance. It must be remembered that in old music fast movements were slower and slow movements faster than in more recent music. Provided one is equipped to judge, one's own feeling as to how fast a piece should be matters more than anything, but even this should be modified by considerations as to how quickly, or slowly, one can play the piece, although if there is a noteworthy difference between the manageable speed and the ideal speed, the piece should be regarded as too difficult for present performance. When the speed of a piece has been decided upon, it should be found on the metronome and marked down on the score for further reference. In the latter stages of rehearsal the whole piece may be played through once or twice with the metronome going: this can be an interesting and salutary experience. You may subsequently have to retard the speed if you find the hall you are playing in is fairly resonant.

Phrasing should now be thought about and the consequential breath marks pencilled in. Breath marks may be made with curved ticks, thus ✓, the size of the tick varying with the size of the breath. Phrasing marks where no breath need be taken should be made with a comma. In consort music, entries should be marked thus ⌐, the thickness of the lines varying with the importance of the entries. Good examples of how to set about this are given in Freda Dinn's *Early Music for Recorders*. Phrasing is dependent upon form. First of all, then, the player must examine the structure of the piece and mark it out into sections. In a sonata these will be statement, development, recapitulation, coda, etc.; in a rondo they will be theme and episodes; in a chaconne each section will be the length of the ground-bass motif; in a fancy the emergence of each new theme to be worked on marks the beginning of each section, generally overlapping the previous section. Even if a piece has no obvious sections into which it can be broken up (double-bars, etc.), a count of the total number of bars and their division by two, three, or four will probably reveal that the piece is, in fact, made up of sections with the same number of bars, usually eight, twelve, or sixteen. Each of these sections should be marked off lightly with a big breath mark, unless a double-bar, a change of key or a long rest makes it superfluous. The 'half-way mark' in each section should next be looked for, and a slightly smaller breath mark lightly pencilled in. When this mathematical process has brought one down to sub-sections of four, six, or eight bars, the shape of the opening theme should be examined, and any modification of sectional breath marks made according to whether it starts on, before, or just after the bar, for generally the position in the bar at which the opening statement starts conditions the phrasing throughout the piece. Next, one must find where the first phrase ends: if this is not evident, it may be revealed by accompanying harmonies, or deduced from the treatment of the phrase later in the piece or in other parts. Once the opening phrase, the germ of the whole piece, has been ascertained, its enunciation should be marked down either by staccato, stress, and slur marks, or by writing over it a pattern of words that will serve as a permanent reminder of its nature. A mixture of mathematics, reasoning by analogy, and good taste will decide the positions of all other breath marks that might be needed. If the music is fast, breath marks may be too far apart to mark all phrasing, so commas should then be used, following the same principles. In preparing for performance, nothing should be left to be dealt with extempore: performances which sound the most spontaneous are those which have been most carefully prepared.

Particular attention must be paid to phrasing in consort music. This is partly because phrases in consort music overlap, partly because their ends are indefinite. Players must decide between themselves when their part becomes less important than someone else's: a good way of preparing a consort piece is to go through it with only the preponderating part playing, the theme being thrown from one player to another. Another approach is for every player to play each part in turn, as if in a round. In Italian and English consort music in particular, one should expect to find breath marks occurring between notes of the same pitch, between two short notes, or between a dotted note and the following short note: phrasing on the beat is more often wrong than right. Players of consort music must depend more on analogy and less on mathematics.

Unless there is a rest, the time taken by breathing must come out of the note before the breath mark. The player must decide exactly how much time he can afford to give to each breath. If he takes too long he might spoil a phrase by cutting the last note short or even endanger a chord: on the other hand if he does not inhale enough air he might spoil the following phrase. He must make allowance for frequent breathing, particularly as under the nervous conditions of performance he will need more breath than in rehearsal. In rehearsal, therefore, the lungs must always feel comparatively full: if they do not, more breath marks should be made in the music. Very long notes constitute a problem. If there are two or more players to a part, arrangements can be made to breathe at different times during the note. Otherwise the player must take a good lungful of air, and, using as low a breath-pressure as possible, hope he lasts out: it is better to break a long note to breathe rather than to peter out in ignominy just before the end. Extended passages of semiquavers are also difficult: if all else fails, a solution is to leave out an occasional note, choosing those that are off the beat, that belong to the chord of the harmony, and that come at the end of decrescendos. The note(s) to be omitted should be ringed and a breath mark put above. If a recorder is playing with strings or keyboard, the other players should know where breaths are being taken and make the necessary allowances in time and phrasing: if the other players are accompanying a recorder, they should breathe with the soloist, lifting their bows or hands at the phrase marks.

Dynamics. If the editor has not already done so, go through the music marking volume indications, working on the principles that no repeat passage is played exactly the same way the second time as the first. In eighteenth-century music echos should be looked for and and marked as such; in consort music each new theme should be announced in such a way that it sounds new – louder, softer, smoother, sprightlier, etc. In sets of variations or chaconnes each section should have its own dynamic level. When more than one player is taking each part, dynamic variations can be achieved by arranging for fewer people to play in the softer passages; if this is done the instructions to the players should be indicated on the music in the preparation stage—'soli' and 'tutti', or something more complicated.

Ornamentation. First, is ornamentation needed at all? If it is, which ornaments are obligatory, which optional? It is well to play only the obligatory ornaments – cadential trills, etc. – at the first playing and to decorate more lavishly for the repeat. Obligatory ornaments might be marked in ink, optional ones in pencil. Every ornament should be marked down and nothing should be deliberately left to improvisation, even though some improvisations might be generated in the heat of performance. The length of appoggiaturas should be noted, especially when two players are trilling together, and for true precision the number of notes in each trill should be settled, unless the trill is a long one. Turned trills should be indicated with a pair of semiquavers showing the turn (the eighteenth-century convention). Pick out and practise the most difficult trills in the piece to bring them up to speed.

Alternative fingerings should be marked with a cross: failure to do so could easily cause a player to be left in panicky indecision with his fingers fluttering ineffectually. Unusual alterntive fingerings might be written out in a memorandum at the bottom of the page. *Vibrato* may be marked with a wavy line, the undulations of which correspond to the amplitude and the 'wave-length' of the vibrato, and *shading* or *slide-fingering* with a downward or upward line, the slope of which denotes the extent of the shading or sliding required. In a *ff* or crescendo passage when shading might be applied over a number of notes the shading line should be extended, sloping farther down as more shading is applied, and the word 'shade' marked in. Editorial phrase-marks or ill-judged slurs that clash with one's own interpretation of the piece may need crossing out or altering. All such markings should be made in pencil, otherwise the technical commentary on the music could easily obscure the notes themselves. Of all the technical apparatus that appears on a thoroughly prepared piece of music, however, nothing matters so much as those two or three guiding adjectives at the beginning of the piece that remind the player how the piece is to be played as a whole.

Memorizing

If you are playing a solo, you may wish to know the music by heart. For a concerto performance, the audience will expect you not to be cut off from them by a stand and music. A good musical memory is a gift which can best be cultivated by achieving close familiarity with the music. Once they have gained confidence, some players play better without the music, because the sound itself reminds them of their own part while they can hear the music in its entirety in their memories. The other parts, especially the bass, remind the soloist of his own part, and in the process, much better ensemble playing is achieved. But a memorized performance, played standing up, is usually associated with music which has a dominant solo part. With the exception of eighteenth-century concertos, some solo sonatas of the period, and pieces for unaccompanied recorder, the recorder's repertoire consists mainly of ensemble music in which all parts are equal. While there are advantages to be gained by all the players knowing the music by heart, it is an unusual accomplishment and not one generally expected by audiences.

Leading

Understanding should be reached before performance on such important details as to who is going to give the lead to start a piece, and who should be watched for the finishing of closing notes (usually but not always the player of the top part); the leader can indicate the tempo of a piece by raising his instrument in time with the beat preceding the start of the music, or he can count out one or more beats with the unused little finger of the left hand. Before beginning the leader should catch the eyes of all the other players to see that they are ready to start, and are not taken unawares with empty lungs right at the beginning of a piece. Music should be marked as to who is giving a lead and an ending, and when two or more players are sharing a piece of music, who is going to turn over the pages. It should be clearly written on the music when repeats are not to be observed.

Final preparation for performance

When the music is prepared, the instrument itself should be looked at. Make sure that the windway is clear from dirt or fluff and that the bore is clean and dry. See that the foot-joint is in the most comfortable position for the little finger. If a keyboard is being used, tune to its pitch with your recorder warmed up before appearing for performance: a consort of recorders should know how far, if at all, each recorder has to be pulled out to be in tune. Recorders should all be warm before performance, and if more than one is to be used a table in a warm place in the room should be available: the head-joints of instruments which are to be most played on should be kept in one's pocket when not being used. If the recorder is warm, clean, and dry, nothing at all need be done onstage before performance, except perhaps to tune with strings. It is possible that the room in which the performance is to take place cannot be as warm as it ought to be, or the atmosphere might be humid: in this case to avoid condensation the player must play as dryly as possible, using a slightly lower breath-pressure and avoiding drinking soon before playing. To take a sea-sickness tablet of the type that lessens the flow of saliva is a good insurance against clogging. Choose to play in the corner of a room from which the sound can be thrown forward, but if a continuo bass or a piano is being used for accompaniment, be near to it; listen to the bass, as it is as important as your own part. For good ensemble, players should be as near to each other as is convenient, and should arrange themselves in the order of the parts they are playing – descant next to treble, treble next to tenor, and so on. Every player should be able to see every other player. Arrange music stands at a height and position so as not to obstruct the audience's view of you; ensure that all the music is on each stand in the correct order; and put a piece of thin, opaque, and neutral-coloured card behind all the music to support it, and to look more handsome from the audience's point of view. Be sure that there is sufficient light for you to see your music.

In the minutes before the actual performance, do four things. Look at the first piece of music you have to play and recall its mood, speed, and phrasing. While you do this, breathe deeply in and out to clear stale air from the lungs and to help you adopt a calm attitude. All the while keep moving the third finger of each hand up and down on their holes – this helps to ensure fluency and independence of finger movement. Remind your feet not to beat time.

All is now ready, and the point is reached where the players, for those few brief seconds, hear in their minds the sound of the opening bars played at their right speed and style, then set in motion that inward rhythm that is to govern the piece, and, picking up its imperturbable beat, begin. Now let the music carry you forward: subject yourself to it, and forget technique.

SELECTED REPERTOIRE

(music to practise)

Earlier editions of *Recorder Technique* contained a historical review of music for (or which could be played on) the recorder. Edgar Hunt's book, and others, now provide such reviews at considerably greater length (see, for example, Colin Sterne's review in Wollitz's *The Recorder Book*, pp. 178–97). This appendix is therefore more of a personal selection of music which an amateur player approaching the advanced stage will wish to explore. The order under each heading embodies up to a point my personal opinion of the importance of each piece – every player should possess his own copies of the music at the top of each list.

'Repertoire' is a concept associated with a soloist's approach to playing, or to public performance. This appendix therefore concentrates on the kind of music which a reader of this book will want to practise privately for possible eventual performance. It deals only briefly with consort and other ensemble music which, though it is demanding in musicianship (particularly phrasing, rhythm, and ensemble) does not require considerable technical expertise.

For much the same reason, no lists are given of music for recorder and voice, though this category includes some of the most attractive and profound music written for the instrument. Hunt's book, and others, show the recorder-player where to look for such music, for example in Bach's cantatas (especially, Nos. 106, 161, and 152), Handel's operas (especially *Acis* and *Rinaldo*), Purcell's theatre music (especially *Dioclesian*) and the Odes (especially that of 1692), and in renaissance music, played, as so often, with mixed voices and instruments. Publishers' catalogues give further guidance, and recorder-players should obtain the catalogues of the publishers referred to in the lists beneath. References to publishers are either given in full, or in abbreviated form, as follows: Ba = Bärenreiter; F = Faber; H = Hinrichsen/Peters; LPM = London Pro Musica;

Mk = Moeck (Universal); OUP = Oxford University Press; S = Schott; U = Universal. References are not always given in the lists where the same music is obtainable from several publishers; judgements then have to be made on the quality of the editing and of the continuo realization. Up to the end of the sixteenth century, composers rarely specified any particular instrumentation for their works – they relied on what was available and on the common sense of performers not to use indoor instruments (like recorders) for outdoor music. Even during the eighteenth century, composers – and publishers with an eye to sales – allowed the substitution of one instrument for another. So it is not 'unauthentic' to play Dowland's *Lachrimae* or Anthony Holborne's 1599 dances (LPM) on five recorders, though the former may more properly be played on viols, and the latter on violins. Recorder-players have every right to regard as their own the instrumental pieces in the *Glogauer Liederbuch*, the Flemish collection in *Liber Fridolini Sichery* (S), the 'Browning' versions by Byrd and others, or even the fancies and suites of Ferrabosco, Lawes, Jenkins, and Locke (S), though this later music (e.g. *Jacobean Consort Music* – Stainer and Bell) has a gravity more proper to the viol consort. Recorders pipe more sweetly in renaissance dance music (LPM) or art-music derived from dance forms, in courtly masque dances (e.g. Adson and Brade), and in lighter-textured fancies (e.g. East, Ward) and canzonas (e.g. Banchieri, Scheidt). Though the recorder yielded pride of place to the cornetto and the violin in early seventeenth-century Italy, and to the transverse flute in early eighteenth-century France, recorder-players of the time would have had no compunction about playing a part in Frescobaldi's canzonas or in Couperin's *Concerts*, perhaps even those of Rameau. Corelli's Op. 5 violin sonatas, including the Follia variations, were published in London for flute (= recorder). Two of Mozart's most popular wind concertos, for flute in D and oboe in C, are virtually the same music (K. 314): why therefore not play Haydn's flute trios or Mozart's flute quartets (e.g. K. 171) on the recorder, occasionally using Procrustean devices to keep them within the range of the treble recorder? Haydn's numerous baryton pieces (of which he published several for flute) sound entrancing on tenor recorder, viola, and cello, with sympathetic string players. The following lists of music, which are in the nature of suggestions for the recorder-player's library, therefore include some music not explicitly written for the recorder.

An asterisk indicates that a 'Music Minus One' record (i.e. recording with the recorder part omitted) is available for practice purposes (agents: Forsyth Bros., 126 Deansgate, Manchester).

Recorder solo

For treble recorder unless otherwise indicated. See in addition the important list of music mentioned in Appendix 2 under 'Ornamentation'.

Preludes and Voluntaries (1708) (S, and Nova).

Fifteen Preludes (S): if you finally learn them, try decorating the repeats. More are provided in the Amadeus edition of Capricen (Quantz).

The Bird Fancyer's Delight (sopranino) (S) 1717: with an excellent introduction by Stanley Godman and an 18th-century mini-tutor.

Telemann *Fantasias* for solo flute (Ba (tenor), S).

J. S. Bach *Partita BWV 1013 for solo flute in A minor*, arr. C minor for recorder (S).

J. S. Bach *Cello suites*, arr. Veilhan (Leduc): easier selection arr. Crepax (Ricordi). Also Suites I–III and movements from violin sonatas and partias arr. Brueggen (Zen-On – U).

Handel *Sonata and Allegro* (from a musical clock) (S).

Frederick the Great of Prussia *Solfeggios* (Sikorski).

Boismortier *Suites* Op. 35 (Flute suites without continuo) (S).

Heberle *Sonata brillante* (for the Austrian *czakan*, an early nineteenth-century variety of descant recorder) (Hännsler).

Lorenz *Variations* (for czaken) (Mk).

Medieval Estampies

Renaissance and later divisions, e.g. Bassano: *Ricercate* (Hännsler)

Country dance tunes (esp. Scottish and Irish) – for velocity and rhythmic lift.

Schubert (or other) songs – to develop a singing tone.

Linde *Modern Exercises* and *Blockflöte virtuos* (S).

Linde *Divertimento* (Mk) – with percussion.

Linde *Fantasien und Scherzi* (S).

McDaniel *The White Tree* (U/Hargail) – attractive exercises in a pastoral vein.

Mönkemeyer *Hohe Schule des Blockflötenspiels* (Mk) – the tunes after each exercise are most engaging, and excellent practice, especially for trills. The present edition substitutes repertoire extracts for these tunes.

Staeps *Virtuose Suite* (S).

Cooke *Serial Theme and Variations* (S).

Schröder *Music for solo treble recorder* (H/Lienau) – 1953, based on tone-rows.

Nobis *Sieben Phasen* (Mk) – 1977; not in the 'very difficult' category, but No. 4 includes harmonics, which require the pitch of the note to be heard in advance.

Alemann *Tropi-Danza* (Barry-Buenos Aires) – 1976; exotic octaving practice.

Owens *Five Compositions* (U) – 1961–6.

Dolci *Nine Ricercari* (H/Heinrichshofen) – 1972.

Sterne *Meadow, Hedge, Cuckoo* (Galaxy) – 1979.

Lechner *Spuren im Sand* (Mk) – 1979; with detailed instructions on technique and interpretation.

Bauer *Bird Pieces* (Breitkopf) – 1964; very difficult, full of special fingerings.

Bauer *Mutazioni* (Breitkopf) – 1962; includes chords, vibrato changes, etc.

Berio *Gesti* (U) – 1966; extremely difficult (even to understand, let alone play).

Further avant-garde music is represented in Schott's *Modern Recorder Series*, ed. Brueggen (Rob du Bois, Andriessen, Shinohara, Geysen, and Casken).

Hunt *Orchestral Studies for Recorder* (S) – in readiness for the moment you are asked to play in a Brandenburg concerto or large-scale music by Britten or Hans Werner Henze.

Duets

It is good practice to record yourself in one part and then play duets with your tape-recorder. The baroque period offers a wealth of music for two recorders or two flutes, much of it intended for amateurs and therefore technically undemanding.

Telemann's duets are outstanding, however, both in their quality and in their technical demands (e.g. one has an exposed high F♯″).

Telemann *Sonata* (2 trebles) in B♭ (*Der Getreue Musikmeister*, 1728) – try this first.

*Telemann *Six Sonatas* (2 trebles) – four in 'Music Minus One': my favourites are those in F and D minor (Nos. 1 and 5).

Telemann *Sonata* for treble and violin.

*Telemann *Six canonic sonatas* (2 trebles).

Telemann *Six sonatas* for two flutes (2 tenors but goes up to E″) – more 'galant' (or 'rococo') in style.

J. B. Loeillet *Sonatas* for two flutes (or 2 trebles), from solo sonatas (*F).

Mattheson *Sonatas* (Opp. 1, 2, 11, and 12) Original duets: No. 11 (with the Chaconne) is good.

Burckhardi *Duets* (S) – Twelve agreeable short pieces.

English baroque duets

The little collection of Thomas Britton, the 'musical coal-man', (1697) is very attractive. For complete sonatas, names to look for in catalogues of recorder music include Daniel Purcell (Henry's brother), William Croft, Godfrey Finger, James Paisible, Robert Valentine, and slightly later and nearer the galant style, Willem de Fesch and G. B. Sammartini.

French baroque duets

This abundant repertoire at first seems rather dull, but played with *bon goût*, i.e. good phrasing, appropriate ornamentation, and a knowledge of French-style articulation and inequality, it comes miraculously to life. Composers include Philibert de la Vigne, Aubert, Boismortier, Hotteterre-le-Romain, Chédeville, Corrette, Naudot (sonatas and 'Babioles'), and Montéclair.

Renaissance duets

Duets, or 'bicinia', were often used for instructional purposes, but some are as finely wrought as multi-part fancies. The following is a short selection from the riches of renaissance two-part instrumental (or vocal) music:

Gibbons *Three Fantasias* (Faber).

Morley *Two-part Canzonets* – including an exciting *Caccia* (hunt).

Lassus *Bicinia* (1601), and *Fantasias* (1577 and 1585).

Gastoldi *Duets* (1598).

Rhaw *Bicinia* (1545).

Sweelinck *Rimes* (1612) (Ba HM. 75).

The LPM catalogue is strong in this area.

Later 18th-century flute duets: other arrangements

The best are six by W. F. Bach, but they range from D to E″. More suitable are duets by Stamitz (Ba), Blavet, Pleyel, Devienne, Geuss (Mk), Bernardo Porta (Presser, U.S.A.), Quantz, and James Hook (S).

Mozart's basset-horn duets go well on recorders (S). Other successful arrangements have been made from C. P. E. Bach, D. Scarlatti, J. S. Bach (two-part inventions) and from viol da gamba duets by F. Couperin and Matthew Locke (Pelikan).

20th-century duets

The following is a selection from a repertoire that is rarely of great technical difficulty:

Tippett *Four Inventions* (descant, treble) (S).

Berkeley *Allegro* (2 trebles) (Boosey and Hawkes).

Cooke *Six Duets* (2 descants) (Mk).

Cole *Suite* (2 descants) (S).

Baines *Variations on an Old Pavan, showing various musical devices* (2 descants) (S).

Horovitz *Ten Duets* (2 descants) (Mills) – exuberant.

Staeps *Reihe Kleiner Duette* (2 trebles) (S) – more difficult.

Staeps *Zu zweien durch den Tonkreis* (2 trebles) (Haslinger).

Chagrin *Six Duets* (treble, tenor) (S).

Gal *Six two-part inventions* (descant, treble) (Haslinger).

Bresgen *Seven Pieces* (treble, tenor) (Mk).

Recorder trios

Music in three parts was common in the medieval period, and in the Renaissance (Henry VIII wrote trios). The repertoire is therefore potentially vast, though never specifically for recorders. Wind trio music is rare in the seventeenth century. There are only three composers who wrote recorder trios in the eighteenth century, all very gratifying:

*Mattheson *Eight trios* (3 trebles) in Op. 1.
Finger *Sonata, Pastorelle* (OUP) and two *Suites* (S) (3 trebles).
Faber *Parties sur les Fleut dous à 3* (trcblc, tenor, bass) (Ba).

Flute trios by Boismortier, Quantz, and Scherer (S) translate fairly well to recorders, and those by James Hook very well. A shadowy Venetian called Carlo Cormier wrote six naïve sonatinas for two recorders and bassoon (Ricordi). There are many excellent arrangements of baroque music for recorder trio.

Modern music for recorder trio includes:

Hindemith *Trio from the Plöner Musiktag* (1932) (descant, treble, treble/tenor) (S).
Müller-Hartmann *Suite* (descant, treble, tenor) (S).
Racine Fricker *Suite* (2 trebles, tenor) (S).
Cooke *Suite* (descant, treble, tenor) (Mk).
Britten *Alpine Suite* (2 descants, treble) (Boosey and Hawkes).

Recorder quartets, quintets, etc.

Again a large repertoire of consort music, polychoral pieces, and baroque arrangements, with an occasional multi-part seventeenth-century work specifying recorders, e.g. Bertali (S), Schmelzer (S), Biber (S). In the twentieth century, recorder players may borrow music for pipes by Vaughan Williams (OUP) and Rubbra, though the latter has written specifically for recorders (e.g. *Notturno*, Lengnick). Other composers in this medium are Arnold Cooke (Mk), Gaston Saux (S), Francis Baines (S), Benjamin Britten (*Scherzo*, Boosey and Hawkes), and Staeps, who writes idiomatic and technically undemanding music for recorder ensemble (e.g. *Seven Flute Dances*, Haslinger, and *The Unicorn's Grace* for twelve or more recorders, Doblinger). Much more avant-garde is Kazimierz Serocki's *Arrangements* for recorder quartet.

Solo sonatas (Recorder and continuo; recorder and piano)

Baroque sonatas are essentially duets between solo and continuo bass (some movements of Handel's sonatas even sound quite nice without the keyboard part). Continuo bass may be gamba, cello, bassoon, or bass recorder, with harpsichord, guitar, lute, or organ to fill out the figured bass and make its own imaginative contribution.

Handel Op. 1 Nos. 2, 4, 7, and 11 (Op. 1 also includes flute, oboe, and violin sonatas).
> G minor – four slowish movements (even the Presto is not very fast). Second movement once goes up to E♭'.
> A minor – the most passionate of the four. Goes up to F″ near the end.
> *C – five movements. The Gavotte is a little tricky.
> *F – the best known, both in its recorder sonata and organ concerto versions.

'Op. 1' was reserved by baroque composers for the presentation of their best early music. Handel's four sonatas are both accessible to the intermediate player and a constant challenge in interpretation and ornamentation to the expert.

Fitzwilliam Sonatas Three (or, as it turns out, two) sonatas mostly put together from other sources. No. 1 is very jolly. No. 3 is less integrated: it contains a movement similar to part of the 'Water Music', and a 'Furioso' which Hunt thinks unsuitable for a 'flûte douce'. All six Handel recorder sonatas are available in an excellent Faber edition.

Telemann *Der getreue Musikmeister* (1728): sonata in F minor (for bassoon or recorder) 'Triste', to the point of despair.
> An impressive and difficult sonata. It quite overshadows the other F minor sonata (S. Ed. 11065).
> F – The one with the top C″, which can easily be avoided so removing any terrors from this sunny piece of music.
> C – Ingratiating slow movements and pyrotechnic (but not all that difficult) fast movements.

Essercizii Musici (1740)
> C – rather similar in style to the earlier sonata in C.
> D minor – impetuous and original. First movement has a double echo.

Two Sonatinas (Musica Rara) – vintage Telemann.

Die Kleine Kammermusik Six Partitas of short, attractive, and quite easy pieces. For violin, flute, or oboe, but suitable for descant recorder.

J. B. Loeillet ('of Ghent') – see articles by Morag Deane in *Recorder and Music* (June and Sept. 1980) 48 recorder sonatas, Opp. 1–4. Op. 1 Nos. 1 and 2 (A minor and D minor) are as good as anywhere to start. The thematic relationship between the movements should be brought out.

Vivaldi (attrib.) *Il Pastor Fido* Five unassuming sonatas followed by No. 6 in G minor, a fine work.

Veracini *Twelve sonatas* for violin or flute (1716) (*Four in MMO) Elegant, yet vigorous.

Bellinzani *Twelve sonatas* (facsimile SPES Florence) Op. 3 No. 12 in D minor (including Follia variations) ed. Lasocki (Nova). Craggy and exciting.

Benedetto Marcello *Twelve sonatas* – 'What oft was thought, but ne'er so well expressed'.

Dieupart *Six sonatas* ed. Bergmann (S): *Six Suites* (Mk) – French-style articulation and ornamentation will bring out the calm beauty of these suites (one of them, in its harpsichord original, was copied out by J. S. Bach).

Barsanti *Sonatas* More imaginative than some other sonatas for the English amateur.

But discover what is best in various sonatas and suites by the following composers:

Pepusch, Finger, Schickhardt (including 24 sonatas in all keys, 1735), Bononcini, G. Sammartini, Mancini, Eccles, Valentine, Galliard, Parcham, Daniel Purcell, and de Fesch. Other names are Dell'Abaco, Bigaglia, Fiocco, Konink, Matteis, and Louis Merci. Two suites by Demoivre are transposed by David Lasocki for bass recorder (U/Hargail). French music of the early eighteenth century is more likely to be for flute than recorder, but instrumentation was very imprecise. There are suites for recorder by Chédeville, Caix d'Hervelois, Bâton, Hotteterre, Marais, La Barre, Lavigne, Philidor, and Senaillé, as well as Couperin's *Le Rossignol en amour* which he said may be played on the flute.

Freedom of instrumentation allows recorder-players to play early baroque sonatas, trio sonatas, and so on by composers such as Rossi, Frescobaldi, Riccio, Castello, Merula, and

Fontana. They are typified by many changes in tempo and exciting cadential flourishes.

Similarly, flute sonatas by later baroque composers such as J. S. Bach, Handel, and Telemann may be appropriated. Some of these appropriations, authentic or otherwise, are very challenging, e.g. Bach's great B minor flute sonata.

Chances of piracy, and some genuine recorder music, e.g. Pugnani (S), may be discovered in the later eighteenth century. Will it ever be quite certain that Gluck's *Dance of the Blessed Spirits* (1762) was intended for the recorder?

20th-century music

My personal favourites among modern pieces for recorder and piano are:

Berkeley *Sonatina* (1940) (S).
Mellers *Sonatina* (1963) (S).
Murrill *Sonata* (1951) (OUP) – technically the least demanding of these three.

Commissions by Manuel Jacobs and Carl Dolmetsch have resulted in the composition of attractive music that differs in style from the German *Gebrauchmusik* sonatas derived from Hindemith (well represented by Genzmer and Burghardt). Composers in the British group include Gordon Jacob, Cyril Scott (*Aubade*, S), Martin Shaw, Norman Fulton (*Scottish Suite*, S), Malcolm Arnold, and Arnold Cooke. Pieces by Edmund Rubbra (Lengnick), Walter Leigh (S), Robin Milford (OUP), and Chris Edmunds (S) are of the pastoral or elegiac sort which well suits the recorder. There is some excellent music by recorder players such as Walter Bergmann, Staeps, Linde, and Colin Hand. Antony Hopkins' descant suite (S) is fairly easy, but very effective.

Trio sonatas

In the baroque period, the trio sonata was as important a medium of expression as was the string quartet in the classical period. The most profound trio sonata is the flute/violin one which J. S. Bach wrote as the centre-piece of his *Musical Offering*: it can, awkwardly, be played on the recorder, but the King of Prussia, for whom it was intended, played the flute. Fortunately the two-flute (or two-gamba) trio sonata may be played on two trebles (H) following the usual baroque custom of raising flute music by a minor third in order to play it on a treble recorder, the lowest note then being F instead of D. Bach's other flute/violin trio sonata (recently published by Hänssler-Verlag) may be played on tenor recorder (though there is a climactic high E″ in the Largo), but the tenor is too soft to go with a modern violin. Use of smaller-toned baroque violins has opened up that other summit of trio sonata writing, Couperin's *Concerts*, to tenor players. Couperin, like other French composers, was very open-minded about instrumentation, and may well have expected variety in performance, as in recent recordings. The upper part does not go below C.

With only a short descent from this Parnassus, we arrive at genuine recorder trio sonatas. Handel's C minor trio sonata (S) is a beautiful work, but being rather low in treble tessitura needs a baroque violin for the second part. The F major (S) does not exploit the lower timbre of the recorder: it is easier, but less rewarding. Faber publish a spirited two-treble trio sonata from Handel's years in Italy. Telemann, however, provides the grateful treble player with a rich vein of silver. Several of the following selected trio sonatas should be in every player's repertoire:

*B flat with harpsichord and continuo (Ba) – thus ideally two harpsichords.
*F with gamba (or viola or cello) (Ba).
*C minor with oboe (H) – from *Essercizii Musici*
A minor with violin (H) – from *Essercizii Musici*.
D minor with violin (treble viol) (S) – with eight bars of semiquavers under one breath in the last movement (*PB* Ex. 14).
*G minor with violin (treble viol) (S).
A minor with violin (S).
D minor with violin (Mk).
F with violin (treble viol) (S: Mk) – happy and undemanding.
F minor with violin (Mk).
A minor with oboe (S: Hänssler) – pyrotechnics.
C for two pairs of recorders with soli and tutti sections (Ba) – in the style of a French suite: the dances have classical names ('Telemann's girl-friends').

F two recorders (Breitkopf & Hartel).
F with horn (H/Noetzel (Pegasus).

Vivaldi wrote three trio sonatas with recorder – a beautiful and not too difficult one in G minor with oboe, a virtuoso trio sonata with bassoon (Musica Rara), a major composition, and one with violin (McGinnis & Marx).

Less demanding trio sonatas or suites for two trebles are by *G. Sammartini, Schickhardt, Corbett, Bononcini, Prowo, Williams (in imitation of birds), Valentine, Paisible, Pez, and surprisingly late, J. C. Schultze (1738–1813). Even more surprising is the existence of a trio sonata for bass recorder, viola, and continuo by C. P. E. Bach, and two trio sonatas by Quantz for recorder, one with violin and the other with transverse flute. The latter inspired Hans-Martin Linde to write a piece with similar instrumentation (1960, S). Many French pieces for two 'dessus' and bass suit recorders, e.g. Marais, La Barre, Dornel, Hotteterre, and Derosier – *La Fuite du Roy d'Angleterre* (OUP), and *Noëls* by de Lalande. Leduc publish for recorders Marais' ingenious *Sonnerie de Ste. Geneviève*, 300 bars on a three-note ground bass. Trio sonatas with violin were also written by Matteis, Fasch, and Pepusch; and with oboe by Lotti, J. Loeillet (of London), *Pepusch, and Fux. Zelenka's two-oboe trio sonatas may be 'borrowed', and recorder-players may even on quite good acoustical grounds appropriate Haydn's lira concerti written for the King of Naples (U/Doblinger). Some seventeenth-century two-violin trio-sonatas, e.g. by Corelli and Purcell, may be tried, although recorder-players may feel well satisfied with Purcell's two great Chaconnes, 'two in one upon a ground' from *Dioclesian*, and 'three parts upon a ground'.

Chamber music with strings, etc.

Vivaldi Lasocki (*The American Recorder*, Fall 1968 and *The Recorder and Music Magazine*, March 1969) lists the following as for recorder: (Pincherle Nos.) P. 77 – a beautiful chamber work with recorder, two violins, and continuo; four for treble, oboe, violin, bassoon, and continuo P. 105, P. 403, P. 207 (Musica Rara) and P. 204 (MR); and P. 81 for treble, oboe, two violins, and continuo (MR).

Telemann Concerto di Camera – treble, 2 violins, and
continuo (S); *Quadro* in G minor – treble, violin, viola, and
continuo (Mk); *Concerto a 4* in A minor – treble, oboe, violin,
and continuo (Mk); *Quartet in F* – treble, oboe, violin, and
continuo (S); and a masterpiece for the unusual combination
of treble recorder, two flutes, and continuo, from the
Tafelmusik of 1733.

Alessandro Scarlatti *Quartettino* – 3 trebles and continuo (H);
Concerto in A minor – treble, 2 violins, and continuo (Mk);
Quartet in F – treble, 2 violins, and continuo (H).

Alessandro Marcello *Concerto di Flauti* – 2 descants and 2 muted
violins, 2 trebles and muted viola, 2 tenors and muted viola,
and continuo (with bass recorder) (Nova).

Fasch *Sonata* – treble, oboe, violin, and continuo (Ba); *Sonata* –
flute, 2 trebles, and continuo – really a trio sonata with the
recorders mainly in thirds (Mk).

Loeillet *Quintet in B minor* – 2 flutes, 2 tenors, and continuo – a
simple work for an unusual combination (Ba).

Schickhardt *Three sonatas* for treble, two oboes/violins and
continuo (H/Lienau); *Sonata, Op. 5 No. 2* – treble, 2 oboes,
gamba/cello/bassoon, and continuo (OUP); *Sonata in A minor*,
Op. 22 No. 6 – 2 trebles, oboe, and continuo (MR); *Six
concertos* – 4 trebles and continuo (Ba).

Finger *Sonata* – 4 trebles, continuo (S, under 'Paisible')

Pepusch *Concerti* Op. 8 Nos. 1, 4, 5, and 6 – 2 trebles, 2 oboes,
and continuo.

Keller *Quintet in D minor* – 2 trebles, 2 oboes, and continuo.

Prowo *Concerta a 6* – 2 trebles, 2 oboes, 2 cellos, and continuo
(Mk).

Telemann *Concerto in A minor* – 2 trebles, 2 oboes 2 violins, and
continuo (S).

Heinichen *Concerto a 8* – 4 trebles, violins, viola, and continuo
(Mk) – treble 1 is 'concertato'.

Hans Gal *Trio* Op. 88 – treble, violin, and cello (Simrock).

Rubbra *Fantasia on a Theme of Machaut* treble, string quartet,
and harpsichord (Lengnick).

Cooke *Quartet* (1964) treble, violin, cello, and piano (S).

Mellers *Eclogue* (1964) treble, violin, cello, and piano
(Novello).

Berkeley *Concertino* (1964) treble, violin, cello, and piano
(Chester).

Martinu *Pastorals* for 2 descants, 2 tenors and bass recorders, 2 violins, clarinet in C, and cello (Ba).

'Piracy' upon chamber music with flute may be tried with Handel, *Concerto a 4* in D minor (S) (now ascribed to Telemann), Boccherini's flute quintets, Haydn's many trios (especially Op. 38 and the *London Trios), *Mozart's flute quartets, and J. C. Bach's quintets.

Concertos

J. S. Bach *Brandenburg Concerto No. 2 in F* (treble, oboe, trumpet, violin, and strings); *Brandenburg Concerto No. 4 in G* (violin, 2 'flauti d'echo', and strings) *or* the same music as *Concerto in F* for harpsichord, 2 trebles (indubitably), and strings.

Vivaldi Lasocki suggests that the Op. 10 concertos (S), as published about 1729–30, were intended for flute or recorder. They include *La Tempesta di Mare* (No. 1), *La Notte* (No. 2), and *Il C(G)ardellino* (No. 3) – try goldfinch chirpings on the sopranino; the last movement is a superb technical exercise in B'–$C\sharp'$ fingerings. Nos. 4 and 5 are relatively simple, but 6 provokes more display. The three flautino concertos can on good grounds be claimed by sopranino players as the ultimate in display, but never to the exclusion of their qualities as music, particularly in the lovely slow movement of P.79. The treble also features in P.440, and four trebles in P.226 (U).

Telemann During the 'dark ages' Telemann's name was kept alive by the *A minor Suite, albeit played on flute instead of the intended recorder. If you cannot manage the *réjouissance*, or the *air à l'italien* (which is both ingratiating and virtuosic), you will find simpler and satisfying music in the *Concertos* in A minor and B flat for two trebles and strings (Ba), in the *Suite for flute pastorelle* (Ba), in the *Chaconne* (2 trebles and strings) from the F minor Suite (S), and in the two Concerti grossi with recorders. The F major concerto (treble, strings, and continuo) (Ba) is fine, but difficult. The E minor concerto (Ba) is for treble, flute, and strings: the last movement is irresistible.

Alessandro Scarlatti *Sinfonie di Concerto Grosso*. Spirited concerted chamber works with no technical difficulties. Nos. 3, 6, 7, 8, 9, 10, 11, and 12 are for flauto, 1 and 5 for two flauti, 2 is for flauto and trumpet, and 4 for flauto and oboe (or violin), all with strings and continuo.

Graupner   *Concerto in F* (treble) (S) – a proper solo concerto, but not all that difficult.

G. Sammartini   *Concerto in F* (descant) (S)

Baston: Woodcock: Babell: Sixth-flute (descant in D) concertos

Cooke   *Concerto* (S) – treble and strings

Jacob   *Suite* (OUP) – treble and strings, but sopranino in the final tarantella.

Concerto soloists lacking an orchestra can try flute music in 'Music Minus One' records, e.g. Bach B minor Suite, Mozart D major concerto, and Andante in C, and concertos by Pergolesi, Quantz, and Frederick the Great.

SELECTED BIBLIOGRAPHY

The following is a list of books, journals, and some music publications which will be of value or interest to recorder-players at the intermediate to advanced stage. It generally excludes books not in English, omitting therefore such important books as Dietz Degen's *Zur Geschichte der Blockflöte in den germanischen Landern* (Bärenreiter, 1939) which has not been translated into English. Many of the books listed are 'standard works' and should be in print. References are under six headings, with some subdivision, and the order within each group reflects to some extent my personal view of the relative importance of each publication to recorder-players; the order should not however be regarded as other than general in this respect, and in any case, several books do not fit particularly well under the headings to which they are allocated. I am indebted to Ruth Davies, Music Librarian at CCAT, and Brian Jordan of Cambridge, for their assistance in compiling this list; and to Brian Jordan also for reading and commenting on the draft of this revised edition of *Recorder Technique*.

1 General books on the recorder
(covering topics in headings below)

Hunt, Edgar *The Recorder and its Music* (Herbert Jenkins, 1962; later editions (Schott/Eulenburg), including paperback, revised) 178 pp.
In revising *Recorder Technique* (first published 1959) I have avoided duplication with Edgar Hunt's book on the assumption that all players who possess my book will also have his. Its eight chapters are on the origin and the history of the recorder and its music (four chapters), the design of recorders, recorder technique (with a historical emphasis), and the recorder in the twentieth century (two chapters).

Peter, Hildemarie *The Recorder – its traditions and its tasks* (Robert

Lienau, 1953; English translation by Stanley Godman, Hinrichsen, 1958) 76pp.
Though badly printed, this book offers more information about recorder playing from a historical point of view than does any other book in so small a compass. It is scholarly and thorough. The book contains valuable summary charts of fingerings and ornamentations.

Linde, Hans-Martin *The Recorder Player's Handbook* (1962; translated by James C. Haden, Schott, 1974) 107 pp.
Another outstanding book (despite some infelicities of translation), with a special emphasis on interpretive problems in recorder music from different periods. Like Edgar Hunt's book, it is well illustrated.

Wollitz, Kenneth *The Recorder Book* (Knopf (USA), 1966; Gollancz, 1982) 259 pp.
This and the preceding three books may reflect national characteristics: Wollitz's book is enthusiastic, engaging, and chatty: it is also well-organised and full of excellent advice on a wide range of aspects of recorder playing. Like Linde, Wollitz emphasizes interpretation; pp. 44–70 on practice contain many valuable hints.
(The above four books all have useful bibliographies.)

Thomson, John M. *Your Book of the Recorder* (Faber, 1968) 75 pp.
An excellent introductory book for the younger reader, with illustrations and short music examples.

Manifold, John S. *The Amorous Flute* (Workers Music Association, London, 1948).
A general handbook for recorder-players and all amateurs of music.

Rigby, F. F. *Playing the Recorder* (Faber, 1958).
An introduction by an amateur enthusiast.

2 On technique
(excluding modern tutors)

Kottick, Edward L. *Tone and Intonation on the Recorder* (McGinnis & Marx, NY; Peters, London, 1974) 27 pp.
There are three chapters – 'Physical Condition of the Recorder',

'Achieving the Optimum Tone' (including photographs of hand and thumb positions), and 'Tuning' (do-it-yourself guide). All recorder-players should think carefully about the effect of different configurations of the oral cavity on tone-quality (and, according to Kottick, intonation); chapter 2 is a good starting-point.

Waitzman, Daniel *The Art of Playing the Recorder* (AMS, New York, 1978) 106 pp.
If one disregards the author's advocacy of the bell-keyed recorder (which he himself recognizes as going out of date), this book is of great value to all recorder-players as one of the most thorough and thoughtful analyses of recorder technique available. Provocative comparisons may be made between Waitzman, Kottick (on tone), and *Recorder Technique*. Even if, as Wollitz says, this book is 'difficult', the points it makes must be tackled by any advanced recorder-player.

Vetter, Michael *Il flauto dolce ed acerbo* Instuctions and Exercises for players of new recorder music (Moeck, Celle, 1969) 87 pp. (German and English, double columns).
Though out of print, this fascinating book must be given a place in this list. It provides nearly three thousand fingerings for recorder notes with different sonorities. It then very ably goes on to deal with various aspects of technique in relation to twentieth-century music. It describes devices and notation employed in avant-garde music. It ends with thirteen pages of hair-raising exercises. (Moeck also publish a booklet by Ursula Schmidt on notation, currently only in German).

Veilhan, Jean-Claude *The Baroque Recorder in 17th- and 18th-century performance and practice – Technique, Performing style, Original fingering charts* (Leduc, Paris, 1980) 70 pp. (French and English, double columns).
Pages 2–20 comprise an excellent compilation from Hotteterre, Freillon-Poncein, and Quantz of instructions for single- and double-tonguing. From then on, the book gives extended musical examples to illustrate transposition (e.g. for voice-flute in D), ornamentation (French and Telemann) and national styles (e.g. the French suite). The book includes a fingering chart consolidated from thirty-one recorder tutors published between 1679 and 1780.

Rose, Arnold *The Singer and the Voice* (Faber, 1962).
Useful for its physiological approach to breath production and control.

Donington, Margaret and Robert *Scales, Arpeggios and Exercises for the Recorder* (OUP, 1961).
For descant/tenor, and treble. This book contains excellent advice on technique and practising.

Although the above does not cover modern recorder tutors, I should like to make special reference to *Treble Recorder Technique* by Alan Davies (Novello, 1983), to Walter van Hauwe's *The Modern Recorder Player* (Vol. 1 Schott, 1984, Vols. II and III to follow), and to Jean-Claude Veilhan's tutors. Quite apart from their precepts, these books contain useful exercises and practice pieces. It is instructive and stimulating to compare where tutors agree and disagree on various points, and how they emphasize different aspects of recorder playing.

Hunt, Edgar *The Bass Recorder* (Schott, 1975).
Bloodworth, Denis *The Bass Recorder Handbook* (Novello, 1986)
Includes great-bass, and sub-contra-bass – fingering, intonation, and music to play.

16th- to 18th-century recorder tutors

Ganassi, Sylvestro *Fontegara* (Venice, 1535; ed. Hildemarie Peter, 1956 and translated by Dorothy Swainson, 1959, Robert Lienau/Hinrichsen) 108 pp.
Recorder-players are fortunate in having so early and so excellent a tutor: it is the starting-point for all historic study of recorder technique. 70 of Ganassi's 90 pages (in the Peter edition) are concerned with divisions – free ornamentation of different intervals in a melodic structure: but the remaining 20 pages are rich with information about breathing ('imitate the expression of the human voice'), fingering, and tonguing. (Ganassi also wrote a viola da gamba tutor, *Regola Rubertina*, 1542/3).

Less comprehensive instructions on recorder playing, but usually including fingering charts, are given by the following (for details see Edgar Hunt's book): Virdung (1511), Agricola (1528 and 1545), Jambe le Fer (1556), Praetorius (1619),

Banister (1681), Salter (1683), and Carr (1686). Freillon-Poncein's tutor for the oboe, recorder, and flageloet (1700) gives detailed instructions for tonguing as well as fingering: it is not yet available in English, except for excerpts in Veilhan (see above), but Minkoff Reprints of Geneva publish a facsimile (1971). Early tutors are often as much concerned with the principles of music as with instrumental technique.

Hotteterre (Jacques Hotteterre-Le-Romain) *Principles of the Flute, Recorder and Oboe* (Paris, 1707); ed. and translated by David Lasocki (Barrie & Rockliff (Barrie & Jenkins), 1968) 88 pp.
The recorder section of Hotteterre's famous tutor opens with an engraving showing the position of the hands on the recorder (would that it were always imitated!) – 'hold *all* your fingers straight'. Instructions are given for normal fingerings (using the then current buttress finger technique with 6 down), trill (*battement*) fingerings, and vibrato (*flattement*) fingerings. Tonguing is dealt with under the flute section ('for flute and other instruments').

Quantz, Johann Joachim *On Playing the Flute* (Berlin, 1752; translated, with introduction and notes, by Edward R. Reilly, Faber, 1966) 368 pp.
Quantz's *Essay of a Method for Playing the Transverse Flute* is largely also applicable to the recorder. It is essential reading for any musician aspiring to play eighteenth-century music.

Other eighteenth-century tutors with instructions for the recorder include Schickhardt (1715), *The Compleat Musick-Master* (1722), *The Modern Musick-Master* (1731– 'Instructions' (8 pp.) 'and Tunes' (40 pp.) ed. Hunt, Schott), Majer (1741), Tans'ur (1746), Sadler (1754), and others up to about 1780.

3 On interpretation, including ornamentation

Donington, Robert *The Interpretation of Early Music* (Faber, 1963, revised 1974) 766 pp.
The standard reference book, and very readable. Its authoritativeness may tempt scholars into disagreement, but lesser mortals may safely regard it as gospel for the time being, especially for the baroque period.

Donington, Robert *A Performer's Guide to Baroque Music* (Faber, 1973) 320 pp.
Shorter by having fewer quotations from original sources, but with greater emphasis on practical applications such as *bel canto* voice production and baroque accidentals. More a guide, less a work of reference. Chapters especially commended are 'Feeling in Baroque music', 'Ornamentation', 'Ornaments', and the whole section on 'Expression' (tempo, rhythm – including the best guide to inequality, punctuation and articulation, and dynamics).

Donington, Robert *Baroque Music: Style and Performance* (Faber, 1982) 206 pp.
A distillation (in paperback) of the previous two books, still with much reference to baroque authorities, but with no material irrelevant to recorder-players. Essential for aspiring players of baroque sonatas, the approach to which requires transparency, incisiveness, order, proportion, impetuosity, and fantasy (p. 171).

Dart, R. Thurston *The Interpretation of Music* (Hutchinson, 1954 and later revised editions) 192 pp.
This book works back from baroque to medieval music. To be read right through by all recorder-players, and re-read at regular intervals.

Veilhan, Jean-Claude *The Rules of Musical Interpretation in the Baroque Era* (Leduc, 1977 translated J. Lambert, 1979) 100 pp.
Written by a recorder-player and with much recorder music among the many extended music examples, this book should be both read and played. It draws mainly upon seventeenth- and eighteenth-century French sources, and Quantz. Together with the second half of Freda Dinn's book, it constitutes the recorder-player's best practical approach to learning about baroque interpretation. The chapter headings are: metrical signs and their characteristics, phrasing, embellishment, the various types of adagio and allegro, and character and tempo of the various movements. Strongly recommended.

Dinn, Freda *Early Music for Recorders: An introduction and guide to its interpretation and history for amateurs* (Schott, 1974) 58 pp.
An essential *gradus ad parnassum*. The second half shows how to play Handel's G minor sonata, a sonata by Thornowitz (1721),

and a suite (descant recorder) by Caix d'Hervelois (1736). The first half shows how to approach consort pieces by Byrd and Holborne.

Bang Mather, Betty *Interpretation of French Music from 1675 to 1775* For woodwind and other performers (McGinnis & Marx/Peters, 1973) 104 pp.
The chapters are 'Rhythmic Inequality', 'Articulation', and 'Ornamentation'. Both short and long examples are given, all within recorder range. A book both to study and play. The section on slurring practices is particularly interesting.

Bang Mather, Betty and Lasocki, David *Free Ornamentation in Woodwind Music 1700–1775* (McGinnis & Marx/Peters, 1976) 158 pp.
A musical anthology of sixty-six original examples of ornamentation.

Ferand, Ernest T. *Improvisation in Nine Centuries of Western Music* (Arno Volk, Cologne, 1961) 164 pp.
Thirty-nine original examples. In this and the previous book the examples are complete movements. Both books have good introductions.

Sadie, Stanley, ed. *The New Grove Dictionary of Music and Musicians* (Macmillan, 1980).
Articles on 'Ornaments' by Robert Donington, and on 'Improvisation, (I) Western' by Imogene Horsley.

Stevens, Denis *Musicology: A Practical Guide* (Macdonald, 1980) 224 pp.
A book for the general reader, concentrating on the interpretation of medieval to early baroque music.

McGee, Timothy J. *Medieval and Renaissance Music – A Performer's Guide* (University of Toronto Press, 1984).

Neumann, Frederick *Ornamentation in Baroque and Post-Baroque Music* (with special emphasis on J. S. Bach) (Princeton, 1978) 630 pp.
A *magnum opus* on the subject, with references to much source material not available elsewhere. Neumann's concern is to banish from performance practice the type of rigidity that results from using in all contexts the basic down-beat patterns of

ornamentation as shown on pp. 116 and 121, without regarding
them as a starting-point from which to develop flexibility, grace,
and rhythmic freedom in embellishment.

Dolmetsch, Arnold   *The Interpretation of the Music of the XVIIth and
XVIIIth Centuries Revealed by Contemporary Evidence* (1915;
reprint of 1946 edition introduced by R. Alec Harman,
Novello, 1969, etc.) 494 pp.
Donington rightly refers to this as 'a pioneering classic'. Even
more so were Dannreuther's two volumes on *Musical Ornamentation* (Novello, 1893–5).

Brown, Howard Mayer   *Embellishing Sixteenth-Century Music*
(OUP, 1976) 79 pp.
A masterly exposition, with music examples, especially of later
sixteenth-century Italian mannerist ornamentation.

Emery, Walter   *Bach's Ornaments* (Novello, 1953, etc.) 164 pp.

Fuller, David   *Mechanical Musical Instruments as a Source for the
Study of Notes Inégales* (Divisions, Cleveland, Ohio, 1979)
(with 7″ record) 20 pp.
A near-equivalent to actual recordings of eighteenth-century
musicians, with some unexpected results.

Music

The following pieces of music, in addition to pieces in the above
books (Bang Mather/Lasocki, Ferand, Veilhan, Dinn, and
Brown), will help recorder-players to know how and where to
ornament:

Telemann   *Methodische Sonaten* (Bärenreiter) Flute Sonatas –
mainly playable on tenor recorder – with a second line for the
slow movements with Telemann's own suggested ornamentation. Some are printed in Bang Mather/Lasocki.
J. S. Bach   *Flute sonatas*, slow movements. Bach was unusual in
that he often wrote out his own ornamentation in full. Bang
Mather/Lasocki reveal two such movements in a simplified
form, as they might have been written, unornamented, by
another baroque composer.

Hotteterre-le-Romain, Jacques   *Duo and Rondeau*, ed. Lefkovitch
(Schott)

Hotteterre-le-Romain, Jacques *Echos*, ed. Bergmann (Schott)

Hotteterre-le-Romain, Jacques *Ornamented Airs and Brunettes*, ed. Lasocki (Nova)

Freillon-Poncein and Hotteterre-le-Romain, Jacques *Préludes* ed. Bang Mather/Lasocki (Faber)

Rippert *Premier receuil de Noëls et Brunettes* ed. Bernolin (Leduc) – these simple duets from 1725 include instruction on where and how to ornament.

Demantius, Christoph *Fifteen Dances 1601* ed. Thomas (London Pro Musica) (includes editor's decorated top part).

Ricercate e passaggi – improvisation and ornamentation 1580–1630 Series published by London Pro Musica edition: includes Dalla Casa, Girolamo (1584) and Bassano, Giovanni (1591) *Divisions on Chansons with original and decorated top parts*. ed. Thomas.

Notari, Angelo *Canzon Passaggiata* (1613) ed. Holman (Nova).

Ortiz, Diego *27 Ricercares* (1553) (Heugel, Paris) – for viola da gamba, but some will go on recorders. Elegant, sequential decoration, an excellent model for improvization.

Van Eyck, Jr. Jacob *Der Fluyten Lust-hof* (1646) ed. Vellekoop (Ixijzet, Amsterdam, three volumes) (also Amadeus).

The Division Recorder (1706 and 1708, two Books) ed. Holman (Shattinger, NY).

4 On the recorder as a historical musical instrument

Welch, Christopher *Lectures on the Recorder in relation to literature* (1911; introduced by Hunt, OUP, 1961).

Manifold, John *The Music in English Drama* (Rockliffe, 1956). On symbolic associations of the recorder.

Baines, Anthony *Woodwind Instruments and their History* (Faber, 1957). No separate chapter, but many references to the recorder. *Nota bene* p. 72: 'Deprived of control through embouchure, a recorder player is obliged to find other methods of keeping the notes steady in pitch throughout the rise and fall of loudness demanded by musical expression. Adjustments with the fingering play a large part in this. The finest virtuosi think ahead all the time to employ the fingering that will best bear the crescendo or diminuendo as the case may be'.

Galpin, Canon Francis W. *Old English Instruments of Music* (Methuen, 1910: fourth edition revised with supplementary notes by Thurston Dart, 1965).

Munrow, David *Instruments of the Middle Ages and Renaissance* (OUP, 1976) with two records (SLS 988).

Original sources

Virdung, Sebastian *Musica Getutscht* (Basel, 1511: facsimile Broude Brothers, NY, 1966).

Agricola, Martin *Musica instrumentalis deudsch* (1529 and 1545) translated by William E. Hetterick in *The American Recorder* (see 'periodicals') and in *The Recorder*, December 1984 and March 1985.

Praetorius, Michael *Syntagma Musicum II De organographia Parts I and II* (Wolfenbüttel, 1619; translated and edited by David Z. Crookes (OUP, 1986).

Mersenne, Marin *Harmonie universelle* (Paris, 1636; facsimile CNRS, Paris, 1975).

5 General background reading

Lloyd, L. S. *Music and Sound* (OUP, 1951).

Dolmetsch, Mabel *The Dances of England and France from 1460 to 1600 with their music and authentic manner of performance* (Routledge & Kegan Paul, 1949, etc. Da Capo paperback, 1976). The companion volume for Spain and Italy is also published by Da Capo.

Roger North on Music (1695–1728) ed. Wilson (Novello, 1959).

Meyer, Ernst H. *English Chamber Music from the Middle Ages to Purcell* (Lawrence & Wishart, 1946; republished 1982 as *Early English Chamber Music . . .*').

Newman, William S. *The Sonata in the Baroque Era* (Norton, NY, 1959, etc.).

Hutchings, A. J. B. *The Baroque Concerto* (Faber, 1961).

Abraham, Gerald (ed.), *The New Oxford History of Music*, Volume VI, *Concert Music 1630–1750* (OUP, 1986).

Harman, Alec and Milner, Anthony *Man and his Music – late renaissance and baroque music* (Barrie & Jenkins, 1962).

Robertson, Alec and Stevens, Denis *The Pelican History of Music* (three volumes).

Brown, Howard M. *Music in the Renaissance* (Prentice-Hall, USA, 1976). Chapter 9 is on instrumental music.

Palisca, Claude V. *Baroque Music* (Prentice-Hall, USA, 1981, etc.).

Blume, Friedrich *Renaissance and Baroque music* (Faber, 1968).

Warburton, Annie O. *Harmony for Schools and Colleges* (Longman, 1938, etc.).

Cole, Hugo *Sounds and Signs: Aspects of Musical Notation* (OUP, 1974). Stimulating, entertaining, and (possibly) prejudice-removing.

6 Periodicals

The Recorder and Music Magazine Incorporating 'The Recorder News' – Journal of the Society of Recorder Players (quarterly) ed. Edgar Hunt (Schott).

The American Recorder Quarterly Publication of the American Recorder Society.

The two main journals in English devoted entirely to the recorder. Both contain articles of great importance to recorder-players. One article (by David Lasocki) in *Recorder and Music* made me change my approach to the first movement of a Handel Sonata, by *reducing* ornamentation.

Early Music (quarterly) ed. Nicholas Kenyon (OUP). A substantial academic journal. Articles of special interest to recorder-players (up to spring, 1986) are:

Ball, Christopher 'Renaissance and baroque recorders: Choosing an instrument' (January 1975).

Best, Terence 'Handel's chamber music' (November 1985) (Handel issue).

Brueggen, Franz 'On the baroque recorder' (April 1974).

Butler, Gregory C. 'The projection of affect in Baroque dance music' (May 1984).

Fitzpatrick, Horace 'The medieval recorder' (October 1975).

Lasocki, David 'Professional Recorder Playing in England 1500–1740:

1500–1640 (January 1982 – a special recorder edition); 1640–1740 (April 1982).

Loretto, Alec 'Recorder modifications: in search of the expressive recorder' (April 1973).
'Recorder modifications 2' (July 1973).
'Recorder modifications 3' (October 1973).

Marissen, Michael 'A Trio in C major for recorder, violin and continuo by J. S. Bach?' (August 1985).

Moeck, Hermann 'Recorders: hand-made and machine-made' (January 1982).

Morgan, Fred 'Making recorders based on historical models' (January 1982).

Neighbour, Oliver 'Orlando Gibbons (1583–1625) The consort music' (July 1983).

Ord-Hume, Arthur W. J. G. 'Ornamentation in mechanical music' (April 1983).

O'Loughlin, Niall 'The recorder in 20th-century music' (January 1982).

Pont, Graham 'Handel and regularization: a third alternative' (November 1985): comment in relation to A minor recorder sonata, Anthony Rowland-Jones (May 1986).

Slim, H. Colin 'Giovani Girolamo Savaldo's *Portrait of a Man with a Recorder*' (August 1985).

Stern, Claudio 'A brief workshop manual for recorders' (July 1979).

Zadro, Michael G. 'Woods used for woodwind instruments since the 16th century': 1 (April 1975); 2 (July 1975).

Various articles on early dance, including Sarabande rhythm (February and May, 1986).

The Galpin Society Journal (especially articles by Eric Halfpenny in Nos. 8, 9, 12 and 13).

The Consort Annual Journal of the Dolmetsch Foundation.

Continuo – a monthly early music magazine (Toronto).

Early Music News – news and reviews, published monthly by the Early Music Centre, London.

Newsletters, etc. of regional Early Music Forums.

INDEX

Names etc. appearing in the Appendices only are not indexed